Embedded Systems

For further volumes:
http://www.springer.com/series/8563

Orlando Moreira • Henk Corporaal

Scheduling Real-Time Streaming Applications onto an Embedded Multiprocessor

Springer

Orlando Moreira
Ericsson
Eindhoven, The Netherlands

Henk Corporaal
Faculty of Electrical Engineering
Eindhoven University of Technology
Eindhoven, The Netherlands

ISSN 2193-0155 ISSN 2193-0163 (electronic)
ISBN 978-3-319-01245-2 ISBN 978-3-319-01246-9 (eBook)
DOI 10.1007/978-3-319-01246-9
Springer Cham Heidelberg New York Dordrecht London

Library of Congress Control Number: 2013953873

Printed on acid-free paper

Springer is part of Springer Science+Business Media (www.springer.com)

Abstract

On a multi-radio baseband system, multiple independent transceivers must share the resources of a multiprocessor while meeting each its own hard real-time requirements. Not all possible combinations of transceivers are known at compile time, so a solution must be found that either allows for independent timing analysis or relies on run-time timing analysis.

This book describes a design flow and software architecture that meets these challenges while enabling features, such as independent transceiver compilation and dynamic loading, and taking into account other challenges, such as ease of programming, efficiency, and ease of validation.

We take data flow as the basic model of computation, as it fits the application domain, and several static variants (such as single-rate, multi-rate and cyclo-static) have been shown to possess strong analytical properties. Traditional temporal analysis of data flow can provide minimum throughput guarantees for a self-timed implementation of data flow. Since transceivers may need to guarantee strictly periodic execution and meet latency requirements, we extend the analysis techniques to show that we can enforce strict periodicity for an actor in the graph; we also provide maximum latency analysis techniques for periodic, sporadic, and bursty sources.

We propose a scheduling strategy and an automatic scheduling flow that enable the simultaneous execution of multiple transceivers with hard real-time requirements described as Single-Rate Data Flow (SRDF) graphs. Each transceiver has its own execution rate and starts and stops independently from other transceivers, at times unknown at compile time, on a multiprocessor. We show how to combine scheduling and mapping decisions with the input application data flow graph to generate a worst-case temporal analysis graph. We propose algorithms to find a mapping per transceiver in the form of clusters of statically ordered actors and a budget for either a Time Division Multiplex (TDM) or Non-Preemptive Non-Blocking Round Robin (NPNBRR) scheduler per cluster per transceiver. The budget is computed such that if the platform can provide it, then the desired minimum throughput and maximum latency of the transceiver are guaranteed, while minimizing the required processing resources. We illustrate the use of these techniques to map a combination of WLAN and TD-SCDMA receivers onto a prototype Software-Defined Radio platform.

The functionality of transceivers for standards with very dynamic behavior – such as WLAN – cannot be conveniently modeled as an SRDF graph, since SRDF is not capable of expressing variations of actor firing rules depending on the values of input data. Because of this, we propose a restricted, customized data flow model of computation, Mode-Controlled Data Flow (MCDF), that can capture the data value-dependent behavior of a transceiver while allowing rigorous temporal analysis and tight resource budgeting. We develop a number of analysis techniques to characterize the temporal behavior of MCDF graphs in terms of maximum latencies and throughput. We also provide an extension to MCDF of our scheduling strategy for SRDF. The capabilities of MCDF are then illustrated with a WLAN 802.11a receiver model.

Having computed budgets for each transceiver, we propose a way to use these budgets for run-time resource mapping and admissibility analysis. During run-time, at transceiver start time, the budget for each cluster of statically ordered actors is allocated by a resource manager to platform resources. The resource manager enforces strict admission control to restrict transceivers from interfering with each other's worst-case temporal behaviors. We propose algorithms adapted from Vector Bin Packing to enable the mapping at start time of transceivers to the multiprocessor architecture, considering also the case where the processors are connected by a network on chip with resource reservation guarantees, in which case we also find routing and resource allocation on the network on chip. In our experiments, our resource allocation algorithms can keep 95 % of the system resources occupied, while suffering from an allocation failure rate of less than 5 %.

An implementation of the framework was carried out on a prototype board. We present performance and memory utilization figures for this implementation, as they provide insights into the costs of adopting our approach. It turns out that the scheduling and synchronization overhead for an unoptimized implementation with no hardware support for synchronization of the framework is 16.3 % of the cycle budget for a WLAN receiver on an EVP processor at 320 MHz. However, this overhead is less than 1 % for mobile standards such as TD-SCDMA or LTE, which have lower rates and thus larger cycle budgets. Considering that clock speeds will increase and that the synchronization primitives can be optimized to exploit the addressing modes available in the EVP, these results are very promising.

Contents

Chapter 1
Setting the Stage

There was a time when the access to a computer was a luxury shared by few. A single computing machine owned by an university or business would service many users, each of which would carefully prepare the programs he or she wished to submit to the machine. After doing the submission, the user would have to wait until the machine could find the time in its busy schedule to process that particular program, and deliver the desired results.

In less than half a century, we find ourselves in a completely different situation. Computers are everywhere. Many common day tasks, like withdrawing money, scheduling an appointment, reading the news or listening to music are handled by computers. We are so used to computers taking care of things for us that we would have severe problems handling our daily routine without them. Besides the conspicuous desktops and laptops, there is a multitude of computers discretely operating inside special-purpose devices. These are the computers in cars, mobile phones, CD/DVD-players, navigation systems, personal digital assistants, and games consoles, to name but a few. In technical circles, it is common to refer to computers included in such devices as "embedded systems."

Some of the most widespread computer systems in our time are those residing in cellular phones. The most recent report from the International Telecommunication Union [44], indicates that, in 2011, there were 85.5 mobile phone subscriptions per 100 inhabitants worldwide, up from 77.2 in 2010, and expected to reach 96.2 in 2013. The same report indicates that in developing countries this percentage has more than tripled since 2005. This incredible level of popularity makes the mobile phone a primary means for deploying new computer-based products and services to the world population at large. Current high-end cellular phones are capable of much more than simple wireless voice communication, including functionalities such as media playing, gaming, personal digital assistant, messaging, satellite navigation, and Internet browsing. However, all this extra functionality still requires connectivity. In fact, a lot of the functionality of most computing machines these days is somehow related with communications. This has to do with a recent trend referred to as "cloud computing" [41, 47], where computing resources are virtualized and

O. Moreira and H. Corporaal, *Scheduling Real-Time Streaming Applications onto an Embedded Multiprocessor*, Embedded Systems 24, DOI 10.1007/978-3-319-01246-9_1,

provided as a service across a network to the user devices. According to this trend, the functionality of consumer devices will tend to become even more dependent on handling communication well. For portable devices, this communication must be wireless.

This book addresses a design problem that frequently arrises in embedded computer systems, and in particular, it focuses on computer systems designed to handle wireless communications as a carrier application for the general methods it describes. The application domain we will focus on is commonly referred to as radio baseband processing. It is an application domain where, in the past, solutions have mainly been designed following a very focused constraint-driven approach, using mostly dedicated hardware, but where the need for an increase in flexibility – the shift from single-radio to multi-radio systems – leads to an increase in unknown factors that will require us to employ more software and adopt more optimality-driven techniques, while trying to preserve as much as possible the essential characteristics of a constraint-driven design.

One of the main requirements of this application is with respect to its timing behavior, and because of that, radio baseband processing is said to be a real-time application. Because of its iterative and data-centric structure, radio baseband processing is also a streaming application.

It is our belief that the techniques presented in this book can be beneficial to the design of any real-time streaming applications, and not only software-defined radio. However, focusing on one area allows for design decisions to be explained with respect to an unambiguous context.

In the reminder of this chapter, we will define what a streaming application is, what a real-time application is, what is involved in radio baseband processing, and how it is implemented in hardware. We will then be able to define the problem that this book addresses.

1.1 Streaming Applications

As in [88], we will define **streaming application** as an application that operates over a long (potentially infinite) sequence of input data items. The data items are fed to the application from some external source, and each data item is processed for a limited time before being discarded. The result of the computation is a long (potentially infinite) sequence of output data items. Most streaming applications are built around signal-processing functions applied on the input data set with little if any control-flow between them.

Besides software-defined radio, examples of streaming applications include other communication protocols, radar tracking, audio and video decoding, audio and video processing, cryptographic kernels, and network processing.

1.2 Real-Time Applications

Many embedded applications are Real-Time (RT) applications. This means that the correctness of the results produced by executing these applications depends not only on the functional correctness of the values produced, but also on the time at which these values are produced. Many computer applications we see in our everyday life have RT requirements. To enumerate but a few, we can refer action computer games, video and audio players, GSM cellular phone transceivers, and controllers for DVD player drives.

The reader may wonder if it is not the case that any computer application is an RT application. A text editor is a good counter-example. Although it is certainly annoying if it takes a long time to re-paginate a document when the user, for example, changes the font type for the whole document, the obtained result – a re-paginated document – is still altogether correct and useful. Per opposition, a first-person shooter game that fails to produce a certain number of frames per second is unplayable; a Wi-Fi modem that fails to acknowledge packet reception to the base station within the time interval specified by the standard will cause the base station to retransmit the same packet over and over again, effectively making communication impossible; a DVD lens focus controller that fails to compute in time the adjustments to the distance between the lens and the disk surface causes the DVD player drive to be incapable of reading the DVD.

RT applications are a heterogeneous bunch. Classifying an application as real-time is normally not enough. It tells us only that the time at which results are produced matters. But what type of timing requirements are there? And how important is it that we meet them?

To put it simply, timing requirements come in two basic types: throughput and latency. If we have a throughput requirement, then the rate at which an iterative application produces results is important, but there may not be any imposition on the time interval between the arrival of an input and the production of an output. If the requirement does prescribe a minimum or maximum time between the arrival of an input data item and production of a related output data item, then it is a latency requirement. A receiver for television broadcast is a good example of an application with a throughput requirement, but without a latency requirement. It is important that the images arrive at a certain rate, in such a way that the illusion of movement is kept for the viewer, at the right pace, but how much time it takes for the image to travel from the broadcast source to the TV screen is not so important (although it is annoying to hear your neighbors celebrate a goal before seeing it 'live' on your own TV). A networked multi-player first-person shooter game is a good example of an application with a latency requirement: the actions of each player must affect the game world in such a way as to seem almost instantaneous to all players.

Temporal requirements can be in the form of a required worst-case timing, a required best-case timing, or both. Sometimes, the worst-case and the best case timing requirements coincide. This is sometimes referred to as an on-time requirement.

We will not discuss on-time requirements or best case timing requirements on this book. We will assume that it is always possible to delay an output event, if that is necessary, and thus focus only on worst-case requirements. In the general case, an RT application can come with several throughput and latency requirements.

Another important classification of RT applications is related with how strict the timing requirement is. There are several classifications of RT applications according to the strictness of the timing requirements. One of the simplest, and still very useful, divides RT applications into two types: soft real-time and hard real-time. Hard real-time applications are those where requirements cannot be infringed under any circumstances, or the results of the computation will be completely useless, and failure may, in the case of a life critical system, have catastrophic consequences. In soft real-time applications, timing requirements can be occasionally disrespected, but the rate of failures must be kept below a certain maximum.

There is also a class of applications where failing to meet the temporal requirements may imperil lives. Such applications are often referred to as critical RT applications. Some authors [12, 51] will only consider hard RT applications the ones that are critical, and prefer to refer to other applications with strict requirements as firm RT. From the experience of the author of this thesis, the latter term is rarely if ever used in industrial settings.

It is important to keep in mind that the classification of a particular application may be in itself a choice that the designer must make. It is a choice where the trade-off is between the delivered quality and the cost of the solution. For instance, because human perception is relatively tolerant of frame loss in a video signal, and since there is a wide variation between the average case and the worst case computational load, it is common to treat video decoding as a soft real-time application, whereas audio decoders, since the variation of computational load is much lower and the effects of losing a sample very noticeable by the user, are more frequently treated as hard RT applications.

1.3 Software-Defined Radio

Software-Defined Radio (SDR) is a streaming application domain where RT behavior is often crucial. The term Software-Defined Radio was used for the first time by Joseph Mitola in 1992 [68]. The idea of SDR is that of relying on programmable processors to implement, by means of software programming, some of the stages of the processing involved in the reception/transmission of a stream of wireless messages, modulated and coded according to a given radio transmission protocol.

Independently of the particular protocol being implemented, digital radio transceivers have a similar flow of data through a number of basic functional blocks, as depicted in Fig. 1.1. They are typically implemented in a number of stages, which include:

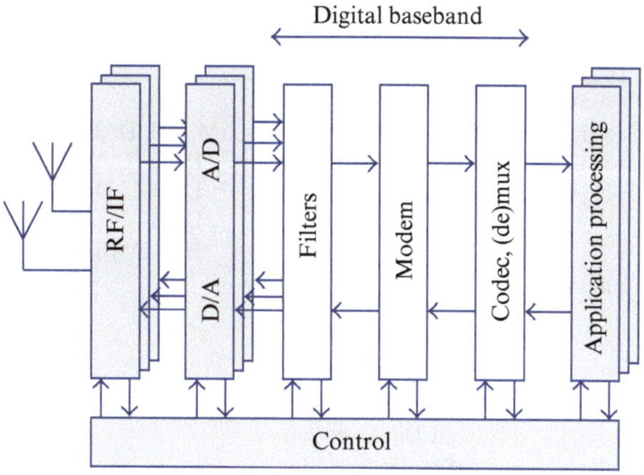

Fig. 1.1 The stages of a radio transceiver

- Radio-frequency filtering and conditioning stage, normally done in the analogue domain;
- Conversion stage, where the analogue signal is sampled and quantized to a digital signal (in the case of the receiver) or converted from digital to analogue (in the case of the transmitter);
- Baseband processing stage, where the digital signal is (de)modulated and (de)coded;
- Application layer, which may include higher layer communication protocols (e.g.: a TCP/IP layer on a WLAN), or simple direct use of the raw received data for some user application, such as voice communication.

In this book, we take software-defined implementations of the real-time baseband processing stage of digital radio transceivers as the carrier application which will inspire and serve as a target for our methods and techniques. The baseband processing stage of a digital radio is completely done in the digital domain. It has strict temporal constraints that imply its treatment as a hard RT application. However, in particular cases, such as Wireless LAN, we will assume that our concern extends to higher level protocols, as these constitute jointly with the baseband a single chain of functional dependencies over which a strict end-to-end real-time requirement is defined.

1.4 Multi-standard Multi-channel Radio

In many devices, there is a huge variety of radio standards that need to be supported. This is both because different standards are developed to handle different types of data transfers, such as audio and video broadcast, two-way telephony, two-way data-link, navigation, and because for each type of communication link there may be several different standards either due to their different technical merits

Table 1.1 Profusion of radio standards

Type	Mobile	Car	Home
Positioning	GPS, Gallileo		GPS, Gallileo
Broadcast	FM, DAB, DVB-T, DVB-H, STiMi	AM, FM, DAB, ISDB-T, ATSC, DVB-C, DVB-T, DVB-HT2,DMB-T	AM, FM, DAB, T-DMB, DRM, XM, Sirius, ISDB-T, DVB-T, DVB-H, HD-Radio, SDARS
Cellular: 3G+	UMTS, HSDPA, HSUPA, MBMS, LTE, TD-SCDMA, CDMA2000, LTE Advance		
Cellular: 2G+	GSM, IS95, IS136, PHS, EDGE, GPRS		
WLAN:	802.11a, 802.11b, 802.11g, 802.11n, WiMax	802.11a, 802.11b, 802.11g, 802.11n, WiMax	802.11a, 802.11b, 802.11g, 802.11n
WPAN	Bluetooth, UWB, NFC	DECT, Bluetooth, UWB, Zigbee	Bluetooth

(range, data rate, latency, vulnerability to noise, etc.) or political and intellectual property-related issues. Different standards are also used for different categories of devices. There are three main market segments for this type of technology: mobile phones, cars, and domestic appliances (home). Table 1.1 gives an overview of standards by market segment and application.

Without an SDR solution, handset makers build their systems by including a dedicated solution for each of the standards they wish to support. This makes the devices inflexible, and does not allow for post-design updates of the functionality.

Modern smartphones must not only support multiple standards, but must also allow for multiple standards to be simultaneously active on a single mobile handset. To illustrate this point, consider an use-case: a hiker using her mobile phone to listen to a digital radio station, via a cordless Bluetooth headset, at the same time using GPS to keep track of her position for posterior plotting of her itinerary, while in the background her phone keeps listening to the cellular network for incoming calls, and an email client periodically checks for incoming emails. This is hardly a far-fetched example, and it requires at least 4 independent radio systems to be active simultaneously.

1.5 Baseband Hardware Architectures

Different types of programs have different patterns of control flow and data manipulation. A filter function, for instance, is typically computed by summing up the results of multiplying 1-to-1 the elements of two long vectors of numeric values;

a decoder, on the other hand, normally involves many bit manipulation operations; a finite-state machine may be dominated by jumps in the program's control flow caused by testing the values resulting from relatively simple arithmetical operations. By designing a processor to excel at one specific type of function, one can achieve better performance, at lower area and power cost. This has led to the development of highly heterogeneous computation platforms, with a set of different programmable cores optimized to handle domain-specific tasks, combined with Application-Specific Integrated Circuits (ASIC) accelerators designed to dramatically speed-up a small set of application-specific functions.

Embedded multiprocessors are not strictly heterogeneous: it often makes sense to employ several cores of the same type. This is done to exploit thread-level parallelism and allow for higher computational capability than would be allowed in a power-efficient manner by increasing the clock speed of a single core (this assuming that an increase in clock speed is possible).

In the case of baseband processing, both academia and industry have independently proposed hardware platforms that follow the general trend of combining homogeneous and heterogeneous multiprocessing, employing multiple vector processors, general-purpose processors and hardware accelerators.

As described in [8], the baseband processing stage can be further split into three sub-stages: Digital Filtering, Modulation/Demodulation (MoDem), and Coding/Decoding (CoDec). These three sub-stages have very different computational characteristics. The Digital Filtering stage has very high computational load (up to five billion operations per second for UMTS), and since the algorithms involved change little from standard to standard, full programmability is not required, and a simple configurable filter is sufficient. The MoDem stage (often referred to as the "inner receiver") is the most diverse among standards. Algorithms in this stage often include heavy processing of vectors and matrix operations. This is typically implemented by means of a vector processor such as the EVP [8], capable of handling multiply-accumulate operations on many input values simultaneously through Single Instruction Multiple Data (SIMD) parallelism. The CoDec stage (also known as the "outer receiver") is more oriented towards manipulation of bits and ordering of data, and, because there is less variety in algorithmic implementation, it is typically handled by a number of ASIC accelerators. The flow control decisions taken by the application are typically handled by general-purpose cores, such as the ones from the ubiquous ARM processor family.

One example of a system architecture for baseband-processing is the MuSIC software-defined baseband chip from Infineon [79], depicted in Fig. 1.2. This platform includes four digital signal processors – Single-Instruction Multiple Data (SIMD) cores, which handle vector operations, a programmable processor, two programmable accelerators, one dedicated to Turbo/Viterbi (de)coding and the other to FIR Filtering, and an array of RF interfaces. Each processor has its own dedicated memory, and a common memory is provided for communication between the cores. The common memory is multi-banked, thus allowing simultaneous accesses from the various bus masters through a multi-layer bus.

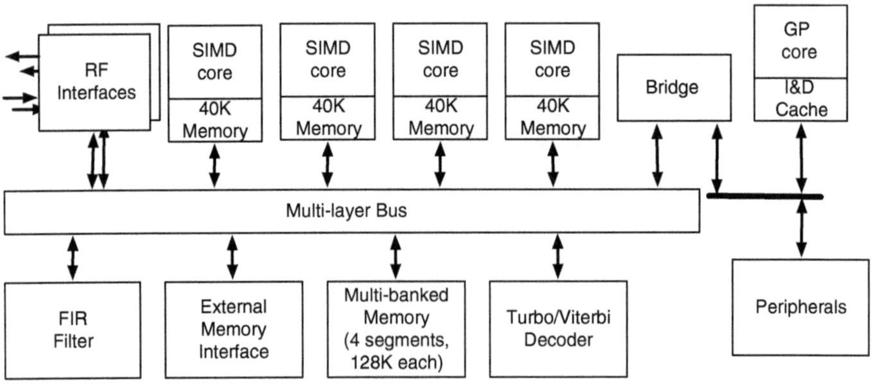

Fig. 1.2 The MuSIC architecture

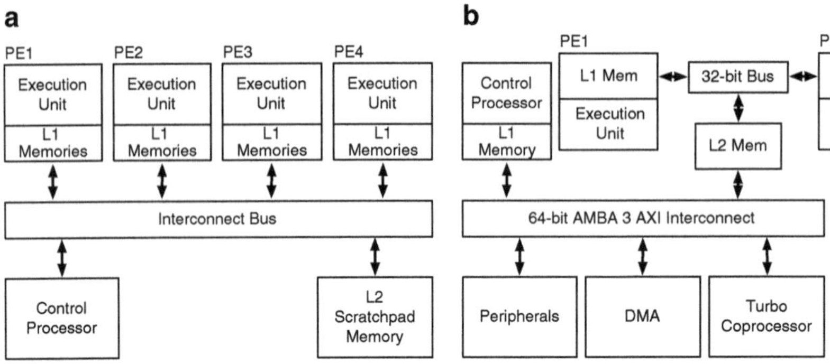

Fig. 1.3 SODA and Scotch

Another architecture proposed for SDR is the SODA architecture [60] from the University of Michigan. It has many similarities with MuSIC. It has recently been redesigned as a commercial ARM Ltd prototype, called Scotch [92]. SODA is depicted in Fig. 1.3a and Scotch in Fig. 1.3b. As MuSIC, SODA employs four SIMD cores, marked as PEs (Processing Elements) in the figure, each with its own local memory, and a control processor. The Scotch prototype introduces changes to the communication and memory hierarchy, including a DMA to handle data transfers between the background memory and the local memories, and hardware acceleration for Turbo decoding.

1.6 Problem Statement

The MuSIC and SODA hardware architectures described in the previous section were designed with single radio operation in mind. As we have seen, most of nowadays applications already require several radio standards to be simultaneously active in the same handset. While it is feasible to deploy one independent programmable platform per standard, it is more than reasonable to expect that, in order to allow maximum flexibility at the lowest cost, radio transceivers will be required to share computation, storage, and communication resources.

The concrete problem this book addresses is as follows: *given a heterogeneous multiprocessor hardware platform, designed for baseband processing, how to manage the hardware resources to allow the simultaneous execution of several hard-real-time applications, in combinations potentially unknown at compilation time, to provide the guarante that each running application will meet its temporal requirements, while using as few resources as possible?*

Most of the challenges in tackling this problem are related with the necessity of providing individual temporal guarantees to each one of the applications. We divide the characteristics that affect the temporal behavior of an application in two main categories: the algorithmic characteristics of the application itself and the characteristics of the execution environment, including both the execution platform and the input stream.

We will discuss these two categories in detail, but to frame that discussion, we need to introduce the notions of determinism and predictability.

1.6.1 Determinism and Predictability

Strictly speaking, a system is said to be **deterministic** if, for a given input, every one of its executions will go through the same exact sequence of states. It is rather useless to rigorously want to apply such a definition to practical computer systems, which must execute in a world whose fundamental physical principles are, as far as we know, non-deterministic [42]. As Henzinger points out in [43], we can often abstract from types of indeterminism that do not affect program execution in any relevant manner: *non-observable indeterminism in the implementation* (i.e. indeterminism, for instance, at the electronic level that does not affect the behavior of logic gates, or in the order of execution of non-dependent instructions on a processor), and *don't care indeterminism* (i.e. indeterminism that only affects parts of the state that we are not interested in). Moreover, Henzinger argues that leaving non-observable determinism out of the scope of our concerns is useful, as it prevents over-specification. For the purposes of this book, we will assume that a program or system is **deterministic** if every time it is executed for a given input sequence, it produces the same output sequence. By defining what we consider part of the output sequence we implicitly define what is left out (the region where "don't care" indeterminism may reside).

Another important related notion is the one of **predictability**. In one of the few attempts we have seen at formalizing the concept, again in [43], predictability is equated to **time-determinism**, that is, a form of determinism where both the input and output sequences are time-stamped. In our opinion, this definition is a good starting point, but it is too strict and misses a couple of important points.

As we discussed in Sect. 1.2, for many RT systems there is no interest in defining the exact times at which outputs must be computed. Instead of this, one wants to guarantee that outputs are produced within certain temporal bounds. Another problem with Henzinger's definition is that it does not take into account whether the system can be analyzed for temporal behavior or not. Saying that a given program is time-deterministic does not imply that general guarantees can be given about its timing behavior, as providing these guarantees could, for instance, require executing the program for all the (potentially infinite) time-stamped input sequences and retrieving all the (potentially infinite) time-stamped output sequences.

We will consider a system **predictable** if there is an algorithm that can provide bounds on the times at which outputs are produced, when a characterization of the timing of the input events is given. This definition still needs qualification regarding two aspects. The first concerns the tractability of the analysis algorithm. If the determination of bounds cannot be done efficiently in time, the system may be predictable, but no temporal guarantees can be computed. Because of this, we will only consider a system **usefully predictable** if it allows us to compute temporal bounds to its outputs within reasonable time. This definition is necessarily ambiguous. We could restrict ourselves to tractable algorithms, but this would leave out techniques such as the one described in [23] for determining the throughput of a timed synchronous data flow graph, which is theoretically intractable, but arguably useful in practice.

The second refinement of the definition is with respect to the tightness of the temporal bounds that one can determine. Tighter temporal bounds allow a much better prediction of behavior than bounds that are less tight. Predictability is therefore not a property that a system either has or not. Although there are systems that are not predictable at all, some systems are more predictable than others.

We will illustrate this point with an example. Consider two single-processor architectures that make use of the same processor core, but with different memory hierarchies. In the first of them, the processor accesses the main memory through a cache. Say that, in this case, a memory read can take anywhere between 2 and 50 processor cycles, depending on whether the access is a cache hit or a cache miss. The read operation in such an architecture is clearly predictable, as we can bound the time it takes to completion of the operation. Now consider the second architecture, where the access to the main memory is direct, and due to the arbitration technique employed, it takes exactly 100 cycles for every access. According to our definition (and according to any intuitive notion of predictability), the second system is more predictable than the first, as the bound on timing behavior is tighter (0 cycles of variance against the 48 cycles of variance in the first case).

But this example also illustrates another important point. It is common for RT practitioners to justify their design decisions in terms of increasing the predictability of the system. This is often a misleading statement. When designing an RT system we are, more often than not, interested in the worst-case temporal behavior of the system. In our example, the worst-case timing of the read operation is much worse for the second architecture (100 cycles) than for the first architecture (50 cycles). The second architecture is more predictable, but the first has a better worst-case behavior. In fact, often we will not be interested in the most predictable system or implementation, but in the one that provides the best worst-case temporal behavior. The reason why a processor's data cache may not be a good option for RT applications is not its unpredictability, but the fact that it does not provide an efficient worst-case temporal behavior.

There is one more thing we wish to say about the relation between predictability and determinism. Functional determinism (i.e. determinism in terms of input and output values, without the added complication of time stamps) is an important enabler of predictable temporal behavior: if the times at which internal actions take place do not influence the outcome of the computation, as it is implied by functional determinism, we may make decisions about scheduling to obtain the desired bounds on the temporal behavior without worrying about affecting the functional behavior of the application. This is an important case of separation of concerns, and a strong reason for preferring functionally deterministic systems over non-deterministic ones.

1.6.2 Algorithm Specification and Temporal Behavior

It is very difficult to infer anything about the timing behavior of an arbitrary concurrent application. One of the most general models for a concurrent application is the multi-threading programming model. It essentially assumes a number of independently executing sequential programs that can read from and write to the same data storage. The problem with such a model, as Edward Lee puts it in [55], is that it is "wildly non-deterministic". Lee describes the work of the programmer of a multi-threaded program as "to prune away non-determinism". Reasoning about the functional and temporal behavior of even the simplest of multi-threaded programs can be extremely challenging. As anedoctal evidence of this, Lee tells the following story:

> A part of the Ptolemy Project experiment was to see whether effective software engineering practices could be developed for an academic research setting. We developed a process that included a code maturity rating system (with four levels, red, yellow, green, and blue), design reviews, code reviews, nightly builds, regression tests, and automated code coverage metrics. The portion of the kernel that ensured a consistent view of the program structure was written in early 2000, design reviewed to yellow, and code reviewed to green. The reviewers included concurrency experts, not just inexperienced graduate students (Christopher Hylands (now Brooks), Bart Kienhuis, John Reekie, and myself were all reviewers). We wrote regression tests that achieved 100 percent code coverage. The nightly build and regression tests ran on a two processor SMP machine, which exhibited different thread behavior than the development machines, which all had a single processor. The Ptolemy II system itself began to be widely used, and every use of the system exercised this code. No problems were observed until the code deadlocked on April 26, 2004, four years later. It is certainly true that our relatively rigorous software engineering practice identified and fixed many concurrency bugs. But the fact that a problem as serious as a deadlock that locked up the system could go undetected for four years despite this practice is alarming.

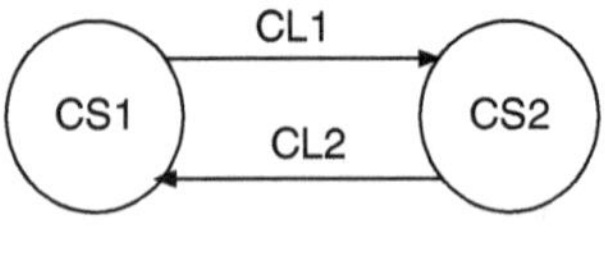

Fig. 1.4 A Kahn process network

> How many more such problems remain? How long do we need test before we can be sure to have discovered all such problems? Regrettably, I have to conclude that testing may never reveal all the problems in nontrivial multi-threaded code.

Lee's story also illustrates how trying to deal with the difficulties associated with programming concurrent applications by developing best coding practices, performing extensive testing, and extensively reviewing the code can fail.

There is a more formal approach to this problem. We can define a restrictive Model of Computation (MoC). A MoC specifies a number of restrictions to programming that guarantee that any program built according to that MoC possesses a number of useful formal properties. Moreover, a well-designed MoC helps the programmer by providing separation of concerns, through the definition of a component model.

A very popular MoC for concurrent applications is the Kahn Process Network (KPN) [48], proposed by Gilles Kahn. A Kahn Process Network consists of a set of deterministic sequential processes (computing stations) executing in parallel and communicating through unbounded FIFO channels (communication lines). Computing stations read and write atomic data tokens from and to the communication lines. The read operation is blocking, meaning that if the reading computing station tries to read from an empty communication line, it will wait and only resume execution once the communication line contains sufficient data tokens to satisfy the read request.

The main formal property that a KPN possesses is determinism: the observed output (the sequences of values written to each communication line) of a KPN is unique for a given input sequence.

Another benefit of KPN is that it provides a separation of concerns between communication and computation, by defining a component model with two types of components, computing stations and communication lines.

The temporal behavior of a KPN application is still difficult to infer. For instance, the problem of determining whether an arbitrary Kahn process network is deadlock-free or not is undecidable [77] (a problem is undecidable if it is impossible to construct an algorithm that leads to a correct answer to the problem for every instance).

Consider the KPN depicted in Fig. 1.4. It consists of two computing stations (CS1 and CS2) and two communication lines (CL1 and CL2). CL1 is written by CS1 and read by CS2, while CL2 is written by CS2 and read by CS1. It is impossible to determine if this graph deadlocks or not without fully characterizing the pattern of communication between CS1 and CS2. For instance, assume that CS2 reads one token from CL1, then writes a token to CL2, then reads again, and writes again, and so on. Assume that CS1 operates in a similar way, first reading from CL2, then

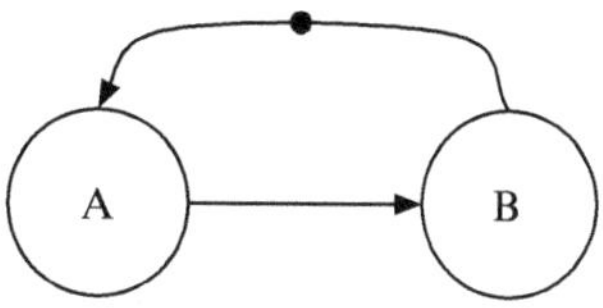

Fig. 1.5 A Single-Rate Data Flow graph

writing to CL1, and so on. In this case, both tasks will immediately block in the first reading attempt, as no data is available. And since both are waiting for the other to produce data, they will both wait forever and the graph will deadlock. If instead one of the tasks is changed such that it first produces and only then consumes, and so on, the graph will never deadlock. This example is very simple. In practice, a deadlock of this type can occur every time there is a chain of cyclic dependencies between computing stations, and the actual occurrence of the deadlock depends on the pattern of reads and writes at each computing station, which may be non-trivial. In our example, if the number of data tokens written or read by any of the tasks would change for each write or read operation, it would be much more difficult, even impossible, to determine whether the graph deadlocks or not.

The fact that it is impossible to devise a single procedure to determine whether an arbitrary KPN is deadlock-free makes it difficult to infer anything about timing behavior in a structured manner.

This problem can be addressed by further restricting the MoC. A popular model for expressing streaming applications is the Single-Rate Data Flow (SRDF) model [80], also known as Homogeneous Data Flow [57]. In SRDF, an application consists of a number of sequential processes (actors), communicating through unbounded FIFO queues (arcs). The difference between KPN and SRDF is that an actor has well-defined activation and data consumption/production rules: an SRDF actor only starts executing when one token is available on each of its input arcs. Once activated, it consumes exactly one token on each of its input edges and produces exactly one token on each of its output edges. Also the initial state of each one of the arcs is part of the specification. Please note that a data flow actor further separates communication from computation. In KPN, although connectivity is independent from computation, the read and write primitives are still expressed as part of the algorithm of the computing stations. In data flow, the read and write behavior of the actor is external to (and defined independently of) its functionality.

Figure 1.5 depicts a simple SRDF graph. It is composed of two actors, A and B, and two arcs, one written by A and read by B and another arc written by B and read by A. The number of data tokens present in an arc at the beginning of the execution must be specified by the programmer. In graphical notation this is commonly represented by black dots on the arcs. In our example, there is one initial token in the arc from B to A. It is easy to determine that our example SRDF graph does not deadlock: at the beginning of execution, only A can execute, since only A has enough input data (the initial token in the arc). A activates, consumes the token in the B to A arc and produces a token in the A to B arc. This ends the activation of A. B can now activate, consumes the token in the A to B arc and produces a token in the

B to A arc. The graph is back at its initial state. Since we have explored all possible states of the graph, it cannot deadlock. It is easy to see that the graph will never deadlock if one or more initial data tokens are present in any the arcs. Also, if more then one initial token is present in these arcs, both A and B can activate at the same time, resulting in parallel execution.

SRDF graphs have a number of interesting formal properties. For instance, it is very simple to determine whether an SRDF graph deadlocks or not [81]. Furthermore, if worst-case execution times are available for each actor activation, then it is possible to compute a fundamental limit to the rate of activation of SRDF nodes [81], sufficient buffer space to guarantee rate-optimal execution and determine rate-constrained static schedules [76]. However, this rich set of formal properties comes at a steep prize: SRDF is a very restrictive MoC in terms of what it can express, requiring every actor to always produce and consume the same amount of data for every activation. Many extensions to SRDF have been proposed to relax these constraints (see, for instance [9, 10, 25, 57]). However, by relaxing constraints, one also loses some of the formal properties. This is essentially the balancing game one plays when selecting or defining a MoC. An important thing one should have in mind is that a MoC should be designed to target a certain application domain. This will ultimately define what formal properties are necessary, and what constraints to what can be expressed are tolerable. As we shall see, our SDR domain is challenging, since it requires more expressivity than SRDF can handle, and requires almost all the formal properties of SRDF. We will further discuss the properties and limitations of SRDF and other data flow MoCs in Chap. 3.

1.6.3 Resource Sharing and Timing Behavior

The Dining Philosophers Problem [18] was originally proposed by Edsger Dijkstra in 1965 as an examination question. It is one of the most famous concurrency problems, and it illustrates some of the difficulties related with resource sharing by non-communicating concurrent processes.

Five philosophers are sitting at a table doing one of two things: eating or thinking. While eating, they are not thinking, and while thinking, they are not eating. The philosophers sit around a round table. Each philosopher has a bowl of spaghetti that always replenishes itself automatically. There is a fork between each pair of adjacent philosophers. Since spaghetti is difficult to eat with just one fork (these are clearly not Italian philosophers!), each philosopher must use two forks to eat. Each philosopher can only use the forks on his immediate left and immediate right. This is depicted in Fig. 1.6.

The problem consists in determining how should we manage the forks (the shared resources) in such a way that none of the philosophers will starve.

The danger of deadlock exists because when a philosopher decides to eat, he needs to obtain two shared resources to proceed. One deadlock state occurs, for instance, when each philosopher is holding his right fork, waiting for the philosopher on his left to release the other fork.

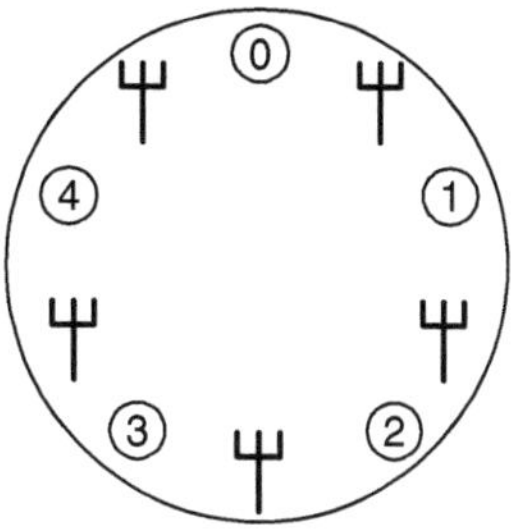

Fig. 1.6 The Dining Philosophers problem

A solution involving a time-out of the wait is not without problems. Let us assume that there is a rule stating that a philosopher puts down a fork after waiting 1 min for the other fork, and waits a minute further before making his next attempt at seizing the forks. Such a scheme eliminates deadlocks, strictly speaking, as the system can always advance to a different state, but it suffers from the problem of livelock: if all five philosophers pick up their left forks at the same time, then all will wait 1 min until releasing their forks, then all will wait for one more minute, then all will pick their left fork once again, and so on.

A simple solution has been proposed for the Dining Philosophers Problem: introduce a waiter. Every time a philosopher wishes to eat, he must ask permission to the waiter before taking any forks. Every time a philosopher finishes eating, he must inform the waiter. The waiter knows where all forks are. He can deny a philosopher his request if four forks are in use at that time. It is still necessary that the waiter grants requests in such a way as to guarantee that all philosophers get their chance to eat (by, for instance, giving priority to new requests from philosophers that had their previous request denied). Moreover, this solution is still dependent on the behavior of each of the philosophers. If one of the philosophers, for instance, never stops eating, or forgets to return the forks to the table when he goes back to thinking, the adjacent philosophers will never be able to eat, because the waiter has no means to enforce good philosopher behavior.

The Dining Philosophers problem shows how it is important to carefully manage sharable resources to avoid (potentially system wide) deadlocks, but it does not directly involve quantitative temporal requirements, only the qualitative requirement that eventually all threads (philosophers) must continue to make progress.

We will now provide a small quantitative example that illustrates the dependence of the timing behavior of a streaming application on another one, when a processing resource is shared between the two.

Consider a processor running an application A. Assume that A is a streaming application that must execute the same function $f()$ within 4 μs of the arrival of each token of input data. Say that the input stream is strictly periodic, with a token arriving every 5 μs, and that we know that $f()$ takes exactly 2 μs to execute. If A has a processor fully dedicated to it, then it is easy to see that A will be able to always meet its temporal requirement: in the initial state, the processor is idle when the first input token arrives, so it can start executing $f()$ immediately; $f()$ then takes 2 μs to execute, producing output within the required 4 μs interval; after this, the processor is back to the idle state; a new input will only arrive 3 μs after $f()$ has finished executing for the first input. At that moment in time, the system is in the same state as at the beginning, and the processing of the second token (and any token after that, for that matter) will just repeat the same sequence of steps as for the first.

Consider now that at a certain point in time, a function g belonging to a second application B starts running on the same processor. Can we still say anything about the temporal behavior of A? This will depend primarily on the way that the sharing of resources is handled. Let us assume that the processor uses a simple non-preemptive round-robin schedule: $f()$ gets to execute, and after $f()$ is done, $g()$ gets to execute and so on. If no input data is available, then the application blocks until data arrives. Can we still say anything about the temporal behavior of A? Well, not without saying something about the temporal behavior of B. If, for instance, $g()$ takes 100 μs to execute for each input data token, it is guaranteed that A will fail to meet its requirement, since every time the processor is busy executing $g()$, as many as $100/5 = 20$ input tokens will arrive for application A, and will have to wait to be processed much more than 4 μs. But if $g()$ takes only 1 μs to execute, any bounding of the temporal behavior of A will still depend on characterizing the activation pattern of $g()$. If, instead, $g()$ receives a new input token every 1 μs, A will always meet its requirement, while if $g()$ receives a new input every 100 μs, A will not be able to keep up with its source, since either the first or the second time that B gets access to the processor, input tokens for A will be accumulating while the processor is blocked waiting for the next input token for $g()$ to arrive. If nothing is know about the pattern of input token arrival for B, then the behavior of A becomes unpredictable.

It is important to keep in mind that the temporal dependence of A on B is introduced by a specific scheduling policy applied to a specific shared resource. If instead we would have chosen to use a time-sliced preemptive scheduler, we would be able to specify the temporal behavior of A by knowing the temporal arrival pattern of its input, the execution time of $f()$, the period of the time slicer and the size of the time slice allocated to A. For instance, if the period of the time slicer is 4 μs and A gets a slice of 3 μs, then A would always be capable of meeting its time requirement independently of the behavior of B. By choosing an appropriate scheduler, we no longer have to worry about the properties of other applications, and our problem becomes how to determine the amount of resources (the size of the slice) that we must allocate to A in order for it to meet its real-time requirements.

This example illustrates another important point: if independence of temporal behavior per application can be guaranteed, then guaranteeing the temporal behavior of each application is much easier. This, again, is a case where we can profit from separation of concerns. Systems where the temporal behavior of each application is completely independent of others are often referred to as **composable** (with respect to temporal behavior) [6, 38, 51]. Multi-processor embedded systems have been proposed [40] that exhibit this property. One pre-requisite of such a system is that either processors only service one application or must be capable of implementing pre-preemptive scheduling policies. As we shall further explain in

Chap. 2, it is one of the original requisites of our software architecture that it must be able to support non-preemptive schedulers, largely because of hardware legacy issues – the SDR multi-radio software architecture should be deployable in existing platforms designed for single radio operation. We will define our ambition in terms of application independence as to provide a platform that is **composable in terms of worst-case temporal behavior**. This means that, although a running application execution time may be affected by other running applications, its worst-case behavior can be inferred independently. Such property is enough to provide the temporal guarantees that our radio applications require. We will come back to this subject in Chap. 2.

1.7 Sketching an Approach

In the previous sections, we have discussed how the algorithm and the execution platform affect the temporal behavior of an application.

Our approach to the design of the software architecture directly addresses these two main sources of unpredictable behavior. We address algorithmic (application intrinsic) unpredictability by employing a strict Model of Computation. We address resource sharing (inter-application) unpredictability by carefully designing the run-time resource management framework to preserve the composability of worst-case temporal behaviors of individual applications.

Our main decision in terms of resource management is that we will use a central resource manager (equivalent to the waiter in the Dining Philosophers example) and a budget-based approach, i.e., applications must explicitly request to a centralized manager for any resources needed to meet their real-time requirements (such as the Dining Philosopher had to ask to the waiter for the forks). A consequence of this is that it must be possible to determine beforehand the amount of resources that an application needs to execute within its real-time requirements.

This need makes data flow MoCs specially interesting. The reason for this is that data flow graphs can be used both as a programming paradigm with strong analytic properties, and as a temporal analysis technique. Starting from a data flow model of the application, one can apply data flow to data flow transformations that represent the temporal behavior caused by specific aspects of the implementation. To illustrate this, consider the SRDF graph represented in Fig. 1.7a. It consists of just two actors. A is a producer and B is a consumer. This graph represents how data flows between A and B without making any assumptions about the timing of operations. Let us now assume we implement the functions that correspond to the activation of A and B in a specific type of processor core. We can now measure the execution time of both functions, and annotate the graph with these. This is depicted in Fig. 1.7b. We can now make predictions about execution times. For instance, for each activation of A and the subsequent activation of B that consumes the token produced by that activation of A, we can conclude that the complete time from start to finish is of $100 + 200 = 300$ processor cycles. This assumes that communication

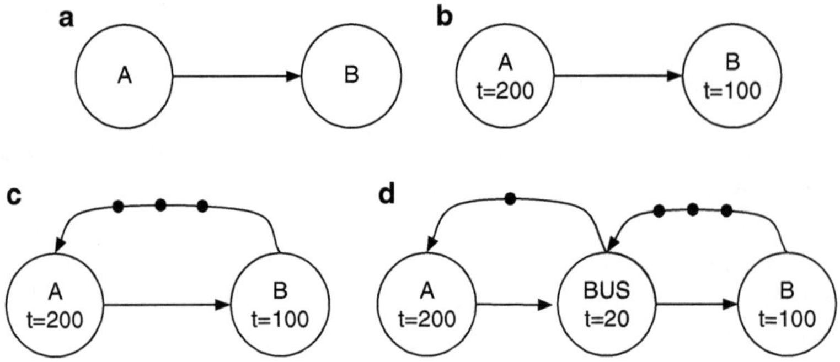

Fig. 1.7 Four different SRDF models of the same application, each refining the assumptions about implementation of the previous one

is instantaneous and that both B and A have full access to the resources of their specific processors. Since no cycles are present, A and B can execute in parallel, the maximum throughput of the system is conditioned by the slowest of the two (we will assume that only one activation of B can be active at a time, which can be modelled in the graph by an edge from B to B with one token). The period of production of outputs (the results of each activation of B) is 200 processor cycles.

Other details of the implementation can be modeled with an SRDF graph. In Fig. 1.7c, we have added an arc from B to A with three initial tokens on it. Such an arc models a limitation to the capacity of the buffer of the arc from A to B: because of the fixed rate of production and consumption of actors, the number of tokens on a cycle is always conserved [81]. Therefore, if B to A starts with 3 tokens and A to B with 0, there can never be more than three data items in the arc from A to B, in any valid state of the system. A buffer space of 3 does not affect the throughput of the system, and the period will still be 200 cycles. However, if this buffer space had been 1, A and B would not be able to execute in parallel, as each would need to wait for the other to transport to each input arc the single token present in the cycle to proceed, and the period would be 300 cycles. More refinements are possible. In Fig. 1.7d, we add to our graph an actor called BUS to represent the latency of communication between A and B through a system bus. Also, the buffering between A and BUS is represented by an arc from BUS to A (A must block while waiting for the transfer operation to start, therefore there is a buffer size of one between A and BUS) and the buffering at the side of the consumer is represented by an arc from B to BUS with 3 initial tokens (the amount of buffer space reserved for the arc at the side of the consumer).

Having our MoC coincide with the timing analysis model, which can also be used to model the attributes of the hardware resources, is a very powerful technique. It allows us to use the same abstraction through the whole programming chain; it allows us to drive the whole process from programming to compiling and scheduling to resource allocation and actual implementation in a correct-by-construction manner.

We also want that as many decisions about resource usage as possible are taken during compilation, in order to alleviate the amount of work that needs to be done at runtime, and allow for the utilization of more complex algorithms. Therefore, our problem involves deciding what should be done by an online resource management framework – essentially an embedded real-time multi-processor multi-tasking operating system – and what should be done by the compilation chain that produces the resource budgets. The solution to our problem is a proposal for a comprehensive software architecture, including an application programming model, a programming/compilation flow, and an operating system.

1.8 Overview of Techniques/Methods Proposed in this Book

To find a solution to our problem that meets the application requirements, we must extend the state of the art in many aspects, including real-time programming and analysis models, real-time analysis techniques, and real-time scheduling and resource management techniques. Specifically, some main topics addressed in this book are the following:

- **Software Architecture Design:** We propose a comprehensive software architecture, including an input language, a programming model, a compilation flow and runtime support, that allows running multiple hard-real-time radio applications, which may start and stop independently of each other, on a multiprocessor system.
- **Data Flow Analysis:** We show that a rate-optimal static periodic schedule can always be found to bound the start times in the self-timed execution of any synchronous data flow graph. We use this property to determine tight worst-case bounds on actor firing times of any self-timed execution of a data flow graph. We establish a property that allows us to characterize the temporal behavior of the self-timed execution of an MRDF graph in the presence of external sources with periodic, sporadic and bursty behavior. We show how these two properties form the theoretical background behing techniques for analyzing latencies and the transient behavior on the self-timed execution of synchronous data flow graphs.
- **Data Flow Modeling:** We describe a customized data flow model – a restriction of Integer-Controlled Data Flow [11], which we call Mode-Controlled Data Flow. It allows deadlock-free execution, static buffer size dimensioning, and quasi-static scheduling, while being able to express the typical data-dependent operation of radio baseband transceivers. We present a technique to analyze the temporal behavior of a Mode-Controlled Data Flow graph across data-dependent transitions. We show the usage of Mode-Controlled Data Flow by using it to model a WLAN 8.11a receiver and a DVB-T receiver.
- **Data Flow Scheduling:** We develop algorithms for static-ordering of actors in MRDF graphs and quasi-static-ordering scheduling of actors in MCDF graphs. We also propose algorithms to determine how to allocate resources to these (quasi-)static-ordered schedules in such a way as to guarantee the hard RT requirements, while minimizing resource usage.

- **Resource Management**: We present algorithms for run-time real-time processor allocation at application start time on a multiprocessor. These algorithms combine processor allocation, network routing, and the allocation of time slices for a multiprocessor system with a network on chip.

1.9 Book Organization

The remainder of this book is organized as follows: in Chap. 2, we make an evaluation of the problem as a whole, derive a set of design requirements from product requirements, and propose an overall structure for the compilation flow and runtime software infrastructure of our solution. In Chap. 3, we review data flow models, and their analytical properties, and introduce mathematical notation for representing data flow graphs. In Chap. 4, we present methods for temporal analysis of the self-timed execution of data flow graphs, and how they can be used to give throughput and bounded latency guarantees. Chapter 5 proposes a method to compute, at compile-time, resource budgets per radio, based on their Single-Rate Data Flow representation. Chapter 6 presents a data flow analysis model, Mode-Controlled Data Flow, that enables the expression of the data-dependent functional behavior of some radio transceivers, while still allowing for strict real-time analysis. This chapter also extends the scheduling techniques presented in Chap. 5 to handle Mode-Controlled Data Flow. Chapter 7 discusses the problem of allocating resources to a transceiver at run-time, based on pre-computed budgets. Chapter 8 describes a prototype implementation of the software architecture, including the resource manager, run-time application support, and compile-time compiler/scheduler. Chapter 9 states our conclusions and suggests future work.

Chapter 2
Software Framework

We present and analyze the requirements of our problem and design a Software Framework (SF) that addresses all of them. We start the chapter by defining terminology and describing our hardware architecture model. We then inventorize the requirements that the SF must address. This is followed by a discussion of the chosen Model of Computation and the overall resource management strategy, including the choice of local schedulers and task synchronization techniques. From this, we identify the main software components of the solution and present an overview of the complete SF, including the compilation chain and the runtime support. We then give a quick overview of components such as the programming language and code generator, which will not be discussed in further detail in this book. We compare our proposal with related work and close the chapter by stating our conclusions.

2.1 Terminology

We will generically refer to a program that processes the baseband layer of a wireless communication channel as a **transceiver**. A transceiver can be instantiated independently of other transceivers. We will refer to a transceiver instance as a **job**. Each transceiver is programmed as a set of concurrently executing **tasks**. Since transceivers are streaming applications, the topology of the communication (i.e. the flow of data and control) between tasks is rather rigid. We refer to the set of tasks and the set of communication channels between them as the **task graph** of the transceiver. We say that a transceiver is **installed** in a device if the device is capable of instantiating the transceiver and executing that instance. We say that a transceiver is **activated** in a device when at least one instance of the transceiver is currently prepared to execute – i.e. all its processes and data structures have been instantiated and the transceiver instance is only waiting for input data – or executing.

O. Moreira and H. Corporaal, *Scheduling Real-Time Streaming Applications onto an Embedded Multiprocessor*, Embedded Systems 24, DOI 10.1007/978-3-319-01246-9_2,

It is also useful to define more precisely what we mean by Software Framework. We consider part of the software framework all the software-related concepts and components that are required in order to program, compile and run applications on the hardware platform. This includes:

1. A model of computation;
2. A programming language;
3. A scheduling and resource management strategy;
4. Application runtime support libraries (e.g., for inter-task communication and synchronization, for job and task instantiation);
5. The runtime resource management and scheduling programs;
6. A temporal analysis model;
7. Temporal analysis algorithms and tools;
8. Compilation/mapping algorithms and tools;
9. Interfaces to the external world.

Our Software Framework must contain all of these components.

2.2 Hardware Architecture

Throughout this book, we will assume as target template architecture a heterogeneous multiprocessor system with many similarities to the architectures presented in Sect. 1.5.

Our multiprocessor system template for baseband decoding, as depicted in Fig. 2.1, includes one or more general-purpose ARM cores, to handle control and generic functionality, one or more vector processors, each an instance of the EVP [8] core, to handle detection, synchronization and demodulation, and one or more Software Codec processors, that take care of the baseband coding and decoding functions. Besides these, the system has an array of RF interfaces, and

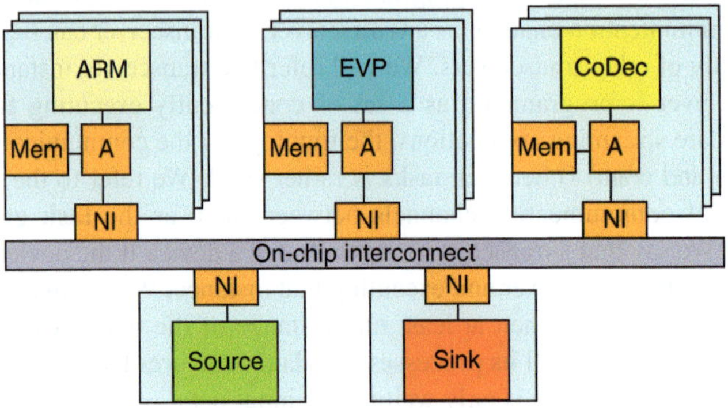

Fig. 2.1 Abstract target architecture template

a memory-mapped interface to the external world. Each processor has its own data and code memory, and caching is not employed. The processors are interconnected by a multi-layer AHB bus, with every processor having both a master and a slave port connected to this multi-layer bus. All masters are connected to all slaves, and therefore communication via posted writes to the consumer's memory is always possible between any two cores. This system template contains the essential characteristics of a programmable platform for baseband. It does not use caches because the improvement on average memory access performance that they provide is not interesting for a hard RT system, that must be designed assuming worst-case timings. Our system employs dedicated local memories per processor core and a simple, fully-connected multi-layer bus with slave-side arbitration to keep the communication flexible, predictable, and easy to handle.

A fully connected multi-layer bus architecture will not scale well with the increase in the number of processors. In later sections of this book, we will extend our architecture with a Network-On-Chip, the Æthereal [30], that is designed to scale with the number of processors, since it does not require full connectivity. The Æthereal Network is particularly interesting for hard RT applications since it allows for the reservation of network resources per channel to provide connections with guaranteed throughput and latency.

2.3 Requirements

In this section we will inventorize a number of important concerns that our SF must address. These will be listed together, categorized only by domain of concern, to avoid over-categorization. It should be obvious from the language employed whether a specific item relates to a strict requirement, a directive, or a nice-to-have property.

1. Application requirements
   (a) **The temporal behavior of jobs must be guaranteed**. A proof must be given that a transceiver meets the RT requirements of the specific radio standard it implements.
   (b) **Jobs must be able to start and stop independently**. During device operation, a request to start executing a job may be randomly issued by an external source, at unknown times. Other jobs may already be active at the time. For instances, a user is transferring information with a wireless LAN and tracking her position with GPS, when a phone call arrives. A new transceiver instance has to be activated, but the other radios should continue to operate without hiccups.
   (c) **The observable behavior of a running job cannot be affected by other jobs**. This relates to our discussion on independence of behavior in Sect. 1.6.3. The SF must assume that a different set of jobs may be present in each device, and that the composition of this set may change dynamically

due to post-deployment upgrades. This implies that worst-case temporal behavior has to be defined independently of the particular combination of transceivers installed. Also, combining jobs in an execution environment should not result in changes to their functional behavior (for instances, we should prevent that a job corrupts the memory space of another job).

(d) **Transceivers must be independently deployable**. This is because post-hardware deployment upgrades (such as new transceivers being downloaded into the device by the end-user) and differentiated user settings (different set of transceivers per device) must be possible. It implies that we either resort to separate compilation of transceivers, or transceivers are compiled in the embedded platform itself.

2. Hardware requirements:

   (a) **The SF must cope with the limitations of the hardware.** Some processing cores cannot support preemption. Because of this, and although preemptive scheduling is desirable in the sense that it allows us to more effectively isolate the behavior of transceivers, the SF must provide both preemptive and non-preemptive scheduling mechanisms. If available, memory virtualization can be used to guarantee that a job cannot corrupt the memory of another job, protecting the system against violations of requirement 1c. Many embedded processors, however, do not have support for memory virtualization or even memory protection. This means that requirement 1c cannot be fully met without rigorously inspecting the code that runs on these processors.

   (b) **Transceivers must be programmed in such a way as to allow efficient mappings of algorithms.** Programming restrictions and primitives provided to the programmer should allow optimized usage of the hardware resources. This includes, for instances, the ability to exploit core intrinsics such as the special-purpose instructions of a vector processor.

3. Software Productivity requirements:

   (a) **Programming transceivers should be easy and intuitive**. The goal of the system is to process baseband radio. The common structure of a baseband processing algorithm should be expressed in a natural and concise way.

   (b) **The SF should make it simple to re-use software components**. Identical software functionality that is required by many transceivers (such as inter-process communication libraries) should be provided as runtime support libraries, and not require repeated independent coding for each transceiver.

   (c) **Mapping should be as much as possible a "push button" process**. Although it is easy for engineers to understand intuitively the major concepts behind temporal analysis of concurrent applications, many of its intricacies are difficult to understand and even more difficult to keep in mind at all times. Moreover, the more human intervention we have in the design flow, the more likely it is that errors are introduced. We want our SF to encapsulate as much as possible the RT analysis techniques, and require as little error-prone manual work as possible.

(d) **Porting transceivers to a different hardware platform should be easy**. Headset makers do not want to be tied to a specific hardware platform, fearing to lose technological leadership of the headset market to a single hardware supplier, as happened to PC makers with respect to Intel in the 1980s/1990s [65]. Because of this, companies like Nokia and Samsung have relied on the strategy of having more than one hardware supplier for modems. To make this strategy work, headset makers must have the ability to quickly adapt software to a different hardware platform. However, in the last few years, the industry has moved more and more to a model where the handset maker expects the provider of the modem hardware to also provide all the modem software, and portability may become less of an issue for the headset maker. In this case, portability becomes an internal issue for the hardware provider, who must internally handle several different hardware architectures, both to the need for different architectures corresponding to different capability classes and price points, the on-going evolution of the hardware architectures, caused by changes and additions to the radio standards.

(e) **Testing and debugging should be simple and manageable**. The complexity of the verification process increases rapidly with the amount of use-cases that must be supported. Our application requires support for a very large number of combinations of transceivers installed in a device, and an even larger number of combinations of active jobs. If a specific bug happens for a specific combination of jobs, it may be very difficult to detect it and localize it. Debugging becomes easier when behavior independence, as per requirement 1c, is guaranteed.

4. Deployment requirements

(a) **It should be possible (and reasonably easy) to port the SF to a different platform**. During the development process of a hardware platform, many architectural decisions may change. The SF should be designed to adapt easily to hardware platform changes. Moreover, it is frequent for subsequent releases of the same embedded architecture to feature more cost-efficient hardware solutions. The SF should remain usable through these iterations. Another aspect of this is that a headset maker is likely to employ different hardware architectures from different hardware suppliers, as explained with respect to requirement 3d. An easily portable SF also makes transceivers easier to port.

(b) **All components of the SF should be useful in isolation, to allow for partial or phased adoption**. One of the main problems in trying to popularize a new framework is that it is often difficult for the potential adopters to accept at once a drastic change to their way of working. If there is a specific advantage for the potential adopter in a single component of the SF, adoption can be phased. The fact that the adopted component offers palpable benefits individually, but larger (synergistic) benefits as part of a whole integrated SF makes it a good introduction to (or a "Trojan horse" for) the complete framework.

2.3.1 *Addressing the Requirements*

Because timing predictability (item 1a) is the most important and strictest of our requirements we will take our major decisions in such a way as to address it, while taking the others into account whenever possible. We have identified in Chap. 1 that the choice of Model of Computation and Resource Management framework are the essential ingredients in defining a predictable system. We will discuss these two choices in the following sections.

2.4 Model of Computation

We opt to use SRDF as our base MoC, with extensions added to fit the application domain and fix the shortcomings of the model, without losing its valuable formal properties. Our extension to SRDF is called Mode-Controlled Data Flow and will be extensively discussed in Chap. 6. As a MoC, Single-Rate Data Flow has many advantages. We already hinted at this in Chap. 1. For now, we will simply list its advantages and disadvantages without further discussion (a more in-depth look at the properties of data flow models will be postponed to Chap. 3). The main advantages of Single-Rate Data Flow are:

- The execution model is data-centric, which follows the natural structure of streaming applications (therefore fitting requirement 3a);
- It separates the specification of the activation rules and data production/consumption patterns of a task (an actor) from its functional specification; this separation of concerns eases the programming effort (requirement 3a) and improves transceiver portability (requirement 3d);
- Strong formal properties: SRDF is functionally deterministic, temporally monotonic and amenable to temporal analysis techniques, which addresses our main requirement, predictability (requirement 1a); moreover, schedules can be statically ordered, reducing runtime scheduling overhead, in accordance to our need for efficient mappings (requirement 2b);
- SRDF can be both employed as a timing analysis model and a programming model; the same abstraction can be used for all stages of the mapping process; a correct-by-construction analysis model can be easily and automatically derived from the specification, and resource allocation decisions can be modeled as SRDF components; this allows for resource budgeting – enabling our strategy to accomplish independent job behavior (requirement 1c) – and automation of the mapping process (requirement 3c)

The main limitations of Single-Rate Data Flow are:

- **Shortcomings of the analysis techniques.** At the moment we started our work, the temporal analysis of SRDF graphs which were not fully-statically scheduled was limited to the analytical determination of a minimum guaranteed throughput on long execution runs. Latency analysis techniques were not available, nor were there any ways to guarantee strictly periodic behavior of output production without applying a fully static schedule. In Chap. 4, we propose extensions to existing analysis techniques to deal with these shortcomings.

- **Limited expressive power.** SRDF actors must have fixed rates of data production/consumption. Many extensions to SRDF have been proposed that improve on this limitation. These, however, tend to lose all the interesting temporal properties. We will give an overview of these data flow models in Chap. 3. In Chap. 6, we will propose Mode-Controlled Data Flow (MCDF), a extension designed to conveniently express SDR applications, without losing the ability to perform temporal analysis.

To use our extension of SRDF as a programming model, we must enforce the usage of SRDF constructs. This calls for a programming language that restricts the user to specify valid data-flow programs. For this effect, we will employ the LIME [54] language. We will give a quick overview of the basic syntax of LIME in Sect. 2.7 of this chapter and extend it to MCDF in Chap. 6.

2.5 Resource Management Options and Choices

The decisions involved in scheduling the resources of a multi-processor system include deciding where to execute tasks (processor assignment) and when to execute tasks (task ordering and task start times). It is common to classify resource management frameworks with respect to when each decision is taken, that is, whether it is done when the application code is generated (compile time) or while the system is in operation (runtime). When a decision is made at compile time, it is referred to as a static decision. When a decision is made at runtime, it is referred to as a dynamic decision. We will discuss options and motivate the choice taken for our SF design in Sect. 2.5.1.

Another decision to be taken is with respect to the way in which task synchronization is achieved across the multi-processor system. This is related with the decision on whether scheduling is handled in a centralized, global fashion, or in a distributed, local way. We will discuss our options in Sect. 2.5.2.

2.5.1 Deciding When to Decide

A survey of the traditional techniques to handle resource management and scheduling of data-flow graphs in a hard-real-time multi-processor system can be found in [81]. These techniques range from fully static to fully dynamic scheduling. Our particular scenario, where jobs are expected to enter and leave the system at any time during operation (stated in requirement 1b), is not considered.

At one end of the range, we have fully static scheduling. In fully static scheduling, the time at which all activities concerning the execution of a program happen is decided at compile time. Combined fully static scheduling of all jobs does not work here because it requires that an exact, unchangeable combination of jobs is known at compile time, fixed execution times for all tasks, fixed communication times for all channels, and a global concept of time across the system.

At the other end of the range, we have fully dynamic scheduling, were all activities are scheduled at runtime. Fully dynamic scheduling is in general possible but impractical. One reason is that it causes high runtime overheads and makes it particularly difficult to give guarantees of temporal behavior. Another reason is that it requires the system to efficiently handle task migration, without affecting the worst-case temporal behavior of jobs.

A traditional way of scheduling a single job is by employing static assignment. In static assignment, processor assignment is done at compile time, but local scheduling is handled at runtime. This does not work in our case, as this strategy cannot take into account that, at the time that a job is activated, an unknown mix of jobs is already running, and using resources. For static assignment to work, the state of the system at the time that the job is started must be known.

An extension of the previous solution that is used in cases where several different combinations of jobs need to be supported is the configuration-based approach (used, for instances, in CPA [84]). For each job combination, a separate optimal static schedule is derived at compile time and stored in a look-up table. During operation, when there is a request to start a job, the runtime system checks which jobs are active and selects the appropriate configuration. This approach has insurmountable problems with respect to our requirements. First, a different configuration has to be stored for each combination of jobs, and therefore the number of configurations may grow exponentially with the number of jobs that must be supported (although in practice there are many job combinations that do not need to be supported). Second, according to our requirements, not all jobs are known at design-time, which means that every time a new job is added to a system, it will force a whole new set of configurations for all jobs to be compiled. Third, as different configurations must assign different resources to each job, it becomes difficult to assure continuity of execution of already running jobs during reconfiguration. If continuity is required, as stated on our requirement 1c, then either there are means for task migration with real-time guarantees,or a configuration must be generated for each transition of a job-mix to another, instead of for each job-mix, which becomes unpractical for even very small sets of jobs.

Our approach is derived from static assignment. However, whereas in traditional static assignment the mapping of tasks to processors is done at compile time, in our case this is done at a specific phase during runtime. At compile time, we calculate independent resource budgets per job. During runtime, two distinct types of temporal phases alternate: reconfiguration phase and steady-state execution phase. During reconfiguration phases, resources are allocated to jobs, while during steady-state phases resource allocation is fixed. Our strategy can also be compared with semi-static techniques [16], where system execution is divided in phases

Table 2.1 Scheduling strategies with respect to the time of decision

	Fully static	Static order	Static assignm.	Configuration based	Our approach	Fully dynamic
Processor assignment	Static	Static	Static	Static per configuration	Semi-dynamic	Dynamic
Task ordering	Static	Static	Static	Static per configuration	Semi-dynamic	Dynamic
Task scheduling	Static	Dynamic	Dynamic	Dynamic	Dynamic	Dynamic

and the resource allocation is redone at the beginning of each phase. But while in semi-static systems phases are typically periodical, in our case reconfiguration phases are triggered when there is a request to start a job.

Our strategy has some similarities with time-multiplexing scheduling strategies such as gang scheduling [19]. Our jobs correspond roughly to a gang in gang scheduling, i.e. a group of tasks with data-dependencies. There is, however, an important difference in granularity: in gang scheduling, time-multiplexing is global, i.e., the temporal slots are uniform across all processors, and synchronized context-switching is required. In our case, time-multiplexing is local to each individual processing element and may be obtained by using any single-processor scheduling mechanism such as Time Division Multiplex (TDM), Round-Robin or even Rate-Monotonic. Our scheduling strategy has several advantages over gang scheduling: it does not require the global synchronization of context switching that gang scheduling needs and which hinders design scalability; it leaves less unused resources because its time-sharing is very fine-grained; and it allows for different local scheduling mechanisms to be used in different processors.

In Table 2.1 we summarize the scheduling strategies we inventorized. From left to right, decisions become more dynamic. The more dynamic the scheduling decisions become, the more difficult it is to give guarantees of timing behavior. With our approach, it is impossible to guarantee that a job can always be started, as it may be the case that there are no resources available at the time the start request arrives or, even in the case that the resources are available, it may be that it is impossible to find the solution: the resource allocation problem, as we shall show in Chap. 7 is NP-complete, even when we disregard memory fragmentation, and, therefore, it is not possible to exhaustively search for all possible resource allocations in efficient time.

This level of dynamicity, however, is necessary because of requirements 1b and 1d. However, it must be noted that these requirements are only with respect to the scheduling between different jobs. Since we know at compile time each job in its entirety, we can alleviate the amount of work that must be done at runtime by statically clustering and ordering tasks within a job. This will be discussed at length in Chaps. 5 and 6. For the moment, we just would like the reader to retain the idea that task ordering is handled dynamically between jobs, but as statically as possible within a job.

2.5.2 *Task Synchronization*

On a typical embedded heterogeneous multiprocessor platform, a general-purpose processor is used to start task executions on all processing elements (accelerators and digital-signal processors), employing the assistance of DMAs to transfer data from one local memory to another. The basic model of execution is essentially single-threaded, with the application control residing singly on the general-purpose processor, acting as a master. All other processing elements are slaves, their tasks being activated by remote procedure calls. Some concurrency can be exploited since the control thread does not need to wait for a remote procedure call to terminate before invoking another procedure call on a different processing element. This strategy may work well when a single, static job is being executed, since not many activities are executed in parallel, and they are all part of the same application. The main limitation of such an approach is that the master is a centralized resource that stands in the critical path of all data transactions and control flow going on in the system. Every time a slave finishes an activity, either it must report it to the master (normally through an interrupt), or the master continuously polls the slaves for events. This does not scale well with an increase in the number of parallel tasks. We opt for distributed data-driven synchronization, where producing tasks inform their consuming tasks about availability of data through a FIFO abstraction, and consuming tasks are activated once data is available. If more than a single task is available at a time, a scheduler local to the processing element decides according to some pre-defined policy what task should execute. Not only does this solve the problem of scalability well, it also fits much better our chosen programming model.

However, not all accelerators are "smart" enough to be used in this way, as their only mode of operation is based on the remote procedure call model, that is, a master indicates to them where input data in memory, were output data must be placed, and starts execution. Such hardware does not support a distributed synchronization model. In such cases, the master/slave interface must be applied. We will say that in such a case the schedule of the slave is hosted by a master, meaning that the master activates a task in the slave and then waits for the task to complete. It is assumed that each slave has a single master, and, since the master waits for task completion, reservation of the host's resources corresponds to reservation of the slave's resources.

2.5.3 *Choice of Local Schedulers*

It is important to choose well the scheduling policies of the local task schedulers. In a budget-based approach, the local schedulers must enable the reservation of resources and give subsequent guarantees of worst-case completion times. In other words, there must be a direct computable relation between a resource allocation and the worst-case completion time of the task. Also, it is necessary that a schedulability check can be performed such that, knowing all the tasks allocated to a processor, their timing requirements and their resource allocation, one can infer whether all tasks will be able to meet their real-time constraints. This requirement rules out popular schedulers, such as preemptive fixed-priority. In a preemptive fixed-priority scheme, each task is assigned a priority and, whenever a task is ready, it will run once no tasks of higher priority are ready. In this scheme, a very high priority task can effectively delay indefinitely the execution of other tasks, depending on the time it becomes active and on how often it is activated. It is therefore impossible to give any guarantees of timely completion to any other tasks in the system without fully characterizing the best and worst case of the activation pattern of the high priority task. Consider for instances that the highest priority task on a given processor, task A, is ready whenever a task B running on a different processor generates data. If B generates data faster, then task A becomes ready more often, and less resources will be available to any other ready task running on the same processor as A, potentially resulting in these other tasks missing their deadlines. This is what is called a timing (or scheduling) anomaly [64], i.e., situations where local worst-cases do not entail a global worst-case.

Although there are many schedulers that fit our requirements, in this book we will consider just two, not so much because of their efficiency, but for their simplicity. In processors that support preemption, we will employ a Time-Division Multiplex (TDM) scheduler, also known as a slicing scheduler. A TDM scheduler has a period P and a list of assigned tasks. Any task A assigned to the scheduler is assigned a time slice S_A. For every period, the scheduler will allow the task to run for a time interval equal to S_A, before preempting it to execute the next task in the list. Knowing the period of the TDM scheduler, the worst-case execution time of a task – the time interval between the moment when the task becomes ready and termination when the task runs alone in the processor without resource sharing – and its allocated slice, one can compute the task's worst-case response time – the time interval between the moment when the task becomes ready and termination when running under supervision of the local scheduler – see [7]. Also, checking if a slice S_A can be mapped to a TDM scheduler with period P amounts to checking if the sum of S_A with all other slices already allocated to the processor is lesser or equal to P.

TDM requires a preemptive scheduler. For processors not capable of implementing a preemptive scheduler, we will use a Non-Preemptive Round Robin (NPRR) scheduler. An NPRR scheduler keeps a list of the tasks allocated to it. It constantly goes through that list, checking at each time if a task is ready. If it is, it allows it to execute until it terminates. If the task is not ready, it checks the next task for readiness, and so on. If we know the worst-case execution time of a task we know that the worst-case response time of the task is equal to the sum of the worst-case execution times of all tasks allocated to the processor.

We will discuss how to compute response times and assign resource budgets to tasks on TDM and NPRR schedulers in Chap. 5.

2.6 Proposed Solution Overview

Our approach addresses the two main sources of unpredictable behavior. We address algorithmic (application intrinsic) unpredictability by employing a strict concurrent Model of Computation (MoC). The restrictions of the MoC are enforced by a programming language, LIME [54].

We address resource sharing (inter-application) unpredictability by designing the runtime resource management strategy to preserve the worst-case temporal behavior of each application. A centralized resource manager enforces admission control and budget-based resource allocation. Jobs (transceiver instances) explicitly request at start time for the resources they need to meet their HRT requirements. Our MoC supports this, by doubling as a temporal analysis model where budgeting decisions can be represented, as already explained in Chap. 1.

We will now give an overview of the proposed Operating System and Programming Flow.

2.6.1 *Operating System*

Figure 2.2 depicts the software stack running on our system. Accelerators are not depicted as their schedule is hosted by the programmable cores. The interface to start/stop baseband jobs is done via the Resource Manager (RM). The RM receives requests from a Central Resource Manager (CRM), as specified in the multi-radio functional architecture proposed in [1]. Upon an admission control request, the RM demands the Configuration Manager (also specified in [1]) for a transceiver budget,

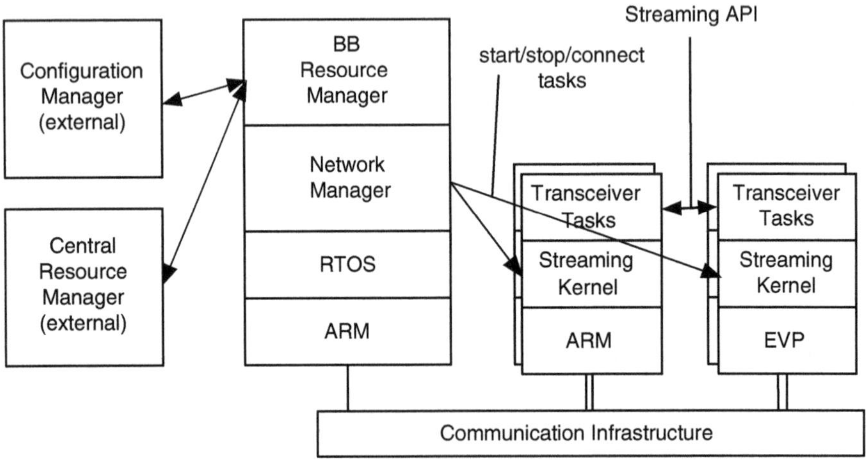

Fig. 2.2 Multi-radio runtime

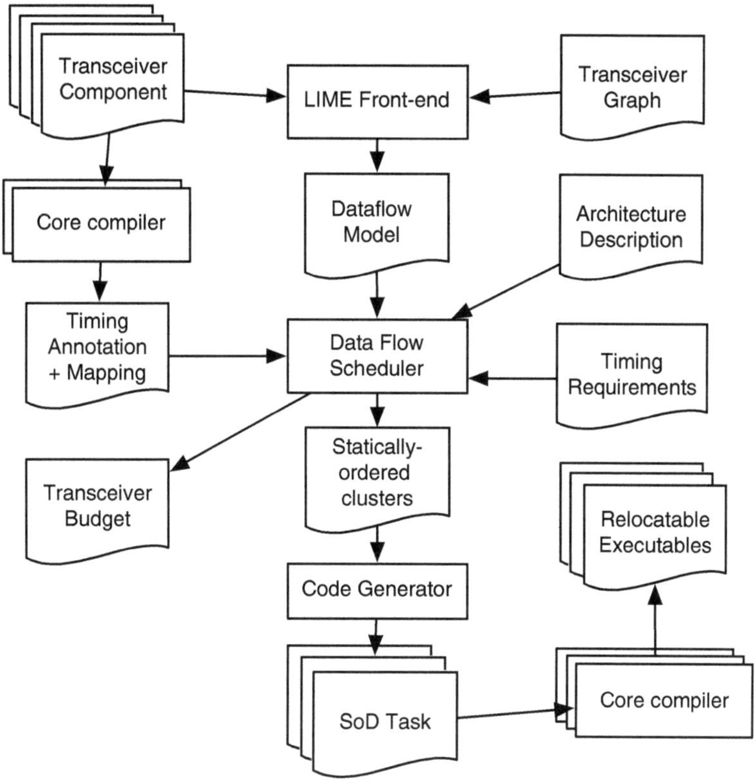

Fig. 2.3 Transceiver programming flow

precomputed by our programming flow (described below). The RM tries to find a mapping that fits the available platform resources. If it finds one, it informs the CRM that the job may be started. The CRM can now instruct the RM to start the job. The RM invokes the Network Manager API to resume all tasks of the job. All cores support a FIFO API, and a local scheduler. In our demonstrator, the Network Manager, local Schedulers and the FIFO API are part of NXP Semiconductors' Sea-of-DPSs (SoD), a lightweight streaming framework for multi-DSP systems. It provides three OS modules: **Streaming Kernels** are single-core task schedulers; the **Streaming API** provides FIFO read/write primitives; the **Network Manager API** creates and destroys tasks and FIFOs on the streaming kernels.

2.6.2 Programming and Mapping Flow

The resource budget of a transceiver is computed at compile time as depicted in Fig. 2.3. The input consists of LIME-compliant C components, implementing DF actors, and the Transceiver Graph, a LIME file that describes how components communicate and synchronize with each other. From these, the front-end extracts a DF model, which is fed to the scheduler. The scheduler also requires a machine description file, describing types and number of cores in the multiprocessor, their local scheduling mechanism, memory segments available, and their sizes. Another input of the scheduler is a partial mapping that provides the scheduler with bindings of LIME components to processor types, and worst-case execution times for all (compiled) components. This file can also be used by the designer to constrain the DF scheduler, by providing maximum values for any budgeted resource. The scheduler also requires the programmer to specify timing requirements. These are minimum throughput and/or maximum latencies.

The scheduler searches for a resource budget for each task and channel that meets the HRT requirements and fits in the multiprocessor resources. It clusters tasks, and computes, per cluster, a quasi-static execution order (explained in Chap. 6), on-line scheduler settings, and, per inter-cluster FIFO, a buffer size requirement. It outputs these as a Transceiver Budget. It also generates a description of how the original software components are ordered within the clusters. The Code Generator uses this to generate a task wrapper per cluster, with calls to the Streaming API for inter-cluster communication and synchronization. Each task wrapper is fed to a core-specific C compiler to generate dynamically loadable object code.

In summary, the DF scheduler solves resource conflicts within a transceiver through clustering and ordering, while the Resource Manager and the Streaming Kernels resolve resource conflicts among transceivers, based on budgets computed at compile time. This division of tasks is done considering that extensive analysis of dependencies cannot be performed efficiently at runtime, and that the set of transceivers is not known at compile time.

2.7 Programming Language

LIME [54] is designed to express applications in multiple data flow MoCs. LIME uses two distinct formats. The graph is specified by an XML schema called GXF. Communication and synchronization are modeled through **ports** attached to LIME components.

A component is a C module, corresponding to a DF actor. The signal-processing algorithms are specified in C99. Processor-specific intrinsics and platform-specific accelerators can be used. This organization isolates platform-specific functionality in some components and supports portability of others.

The signature of the top-level C function in a module declares the ports and firing conditions of the actor. Each argument corresponds to a port and it is declared as an array of the type of data received. The variable name provides the name of the port; the *array size* specifier represents a production/consumption rate, while the presence/absence of the `const` type modifier indicates whether the port is input or output. For example, a Gain actor that outputs the value of the input sample multiplied by 5 can be written like this:

```
void Gain (const int in[1], int out[1])
{out[0]=in[0]*5;}
```

Since arguments are arrays and C arrays are pointers, the Code Generator is free to decide how to implement data acquisition and pass pointers to the C function. It can use any FIFO implementation (blocking/non-blocking, in-place/double buffered, with/without synchronization), or even replace FIFOs for local buffers when combined with clustering. This enables automatic code generation of all the communication primitives, allowing re-use and improving efficiency of the implementation. It also makes the code easier to port, as different, hardware-specific implementations of FIFO communication can be generated by simply re-writing the Code Generator.

Consider the SRDF graph in Fig. 1.7a. Assume the prototypes of A and B are

```
void A (int out[1]);
void B(const int in[1]);
```

If the DF scheduler creates separate tasks for A and B, the generated task wrappers, using the Streaming API, are:

```
void taskA(channel fifo){
int *p;
if !(p = checkWrite(fifo)) return;
A(p);
Write(fifo ,p);
updateWrite(fifo);}

void taskB(channel fifo){
int i;
if !(checkRead(fifo)) return;
i=Read(fifo);
B(&i);
updateRead(fifo);}
```

Instead, A and B can be clustered and statically ordered and the Code Generator instructed to produce a single task wrapper:

```
void taskAB(){int i; A(&i); B(&i);}
```

In this example, clustering saves 1 task wrapper, 6 API calls per firing, and 1 FIFO, and reduces the number of task to be scheduled (and therefore the scheduling overhead) by 1 task.

The output port of actor A can be connected to the input port of B by a GXF statement equivalent to "FIFO `f1` connects port `A.out` to port `B.in`".

Besides FIFOs, LIME allows actors to connect to initialization edges, for delay initialization; and to state edges, to carry actor state across firings (actors are otherwise stateless).

LIME can express several DF flavors. Strict compliance to MCDF is verified by checking the construction rules described in Chap. 6. Since actors are written in C, processor intrinsics and core-specific C compiler optimizations can be exploited. This cannot be done for dedicated languages like StreamIt [88]. As it does not explicitly code communication with API calls, nor allows interleaving of communication with computation, LIME enables automatic extraction of MCDF analysis models, and generation of tasks per cluster with platform-specific communication. This is not possible for API-based extensions of C such as MPI [36] (although MPI could be used as target for code generation instead of SoD).

2.8 Related Work

There have been a few programming models and software architectures proposed for SDR.

One software architecture that is widely used for SDR for military applications is the Software Components Architecture (SCA) [46], proposed by the the United States Army. Its uses an object-oriented programming model. It allows allocation of resources to a transceiver at start time. It does not require a disciplined use of concurrency, relying on general-purpose multi-threading. Consequently, temporal analysis is not possible, and timing must be verified through exhaustive simulation of use-cases.

Another software architecture for SDR is proposed in [35]. It uses a dependency graph for compile time scheduling. The graph is drawn manually by the designer, and may not conform to the actual implementation, which is programmed using multi-threading primitives. Real-time analysis is not performed, and validation is done by extensive simulation. For runtime support, it employs a multi-threaded operating system, and time-sliced schedulers. This work does not address multi-radio. The authors stress that verification of the multi-threaded program is the most complex, time consuming part of the design.

In [61], a programming language and flow that handle single-radio are presented. HRT behavior is specified by a synchronous language. Synchronous languages have difficulties in dealing with multiple independent applications: separate compilation, if not impossible, is difficult; pipelined behavior and the effect on timing of runtime schedulers is not handled. The argument against using synchronous languages in this context is rather long, mostly because many attempts have been made recently to partially fix the shortcomings of the approach; in [90] the subject is discussed in more detail.

2.9 Conclusions

HRT multi-radio requires a budgeted, admission-controlled approach to resource management, that isolates the worst-case behavior of transceivers. Furthermore, a choice must be made about what part of resource allocation is done at compile time, and which part of resource allocation must be done at run time.

HRT guarantees require a Model of Computation amenable to temporal analysis. We depart from Single-Rate Data flow which we will extend to Mode-Controlled

Data Flow, which is customized to express the limited data dependent behavior of transceivers. The MoC must provide a temporal analysis model that can double as a programming model to express the limited concurrent behavior of transceivers and hardware budgeting decisions. DF also acts as a component model that isolates computation from communication, making task interfaces explicit, and improving code portability. With LIME, the programmer can write highly optimized sequential code for each task, and specify communication at an abstract level. This reduces concurrency coding errors, and saves programming effort, as calls to communication APIs are automatically generated at compile time.

The importance of combining a strict MoC with a programming language that enforces it to provide correct-by-construction models for temporal analysis and scheduling cannot be overemphasized. We previously tried direct usage of concurrent libraries such as SoD's by the programmer, with manual extraction of a DF model by an expert. This was a long and hard process, riddled with errors and wrong assumptions. Furthermore, as mapping decisions were manually implemented, guarantees of a correct final implementation could not be given, and extensive testing was required. The approach presented in this book greatly reduces the verification effort and enables HRT guarantees.

Chapter 3
Data Flow Computation Models

In the previous chapters, we justified our choice for data flow as the basis of a model of computation for embedded real-time streaming applications running on a multiprocessor. There are many flavors of data flow. The ones that are interesting for our stated goals are mostly the variants that exhibit behavior which is independent of data values, because of their analytical properties and the variants with deterministic, data value dependent behavior, because of their expressivity. In this chapter, we present notation for data flow models that we will use throughout the book, and the properties of several data flow computation models that are relevant to our stated goal. This is reference material. It can, for the most, be found elsewhere in the literature [10, 57, 80, 81].

3.1 Graphs

A **directed graph** G is an ordered pair $G = (V, E)$, where V is the set of **vertexes** or **nodes** and E is the set of **edges** or **arcs**. Each edge is an ordered pair (i, j) where $i, j \in V$. If $e = (i, j) \in E$, we say that e is **directed** from i to j. i is said to be the **source node** of e and j the **sink node** of e. We also denote the source and sink nodes of e as $src(e)$ and $snk(e)$, respectively.

3.1.1 Paths and Cycles in a Graph

A **path** in a directed graph is a finite, nonempty sequence $(e_1, e_2, \ldots, e_n)$ of edges such that $snk(e_i) = src(e_{i+1})$, for $i = 1, 2, \ldots, n-1$. We say that path $(e_1, e_2, \ldots, e_n)$ is **directed from** $src(e_1)$ to $snk(e_n)$; we also say that this path **traverses** $src(e_1), src(e_2), \ldots src(e_n)$ and $snk(e_n)$; the path is **simple** if each node is only traversed once, that is $src(e_1), src(e_2), \ldots src(e_n), snk(e_n)$ are

O. Moreira and H. Corporaal, *Scheduling Real-Time Streaming Applications onto an Embedded Multiprocessor*, Embedded Systems 24, DOI 10.1007/978-3-319-01246-9_3,

all distinct; the path is a **circuit** if it contains edges e_k and e_{k+m} such that $src(e_k) = snk(e_{k+m}), m \geq 0$; a **cycle** is a path such that the subsequence $(e_1, e_2, \ldots, e_{n-1})$ is a simple path and $src(e_1) = snk(e_n)$.

3.2 Multi-Rate Data Flow Graphs

A Multi-Rate Data Flow (MRDF) graph – also known as Synchronous Data Flow [57, 81] – is a directed graph, where nodes are referred to as actors, and represent time consuming entities, and edges are referred to as arcs and represent FIFO queues that direct values from the output of an actor to the input of another. Data is transported in discrete chunks, referred to as **tokens**. When an actor is activated by data availability it is said to be **fired**. The condition that must be satisfied for an actor to fire is called the **firing rule** of the actor. MRDF requires strict firing rules: the number of tokens produced (consumed) by the actor on each output (input) edge per firing is fixed and known at compile time. During an execution of a data flow graph, all the actors may fire a potentially infinite number of times.

In a timed MRDF graph, actors have a valuation $t : V \rightarrow \mathbb{N}_0$; $t(i)$ is the execution time of i. Arcs have a valuation $d : E \rightarrow \mathbb{N}_0$; $d(i, j)$ is called the delay of arc (i, j) and represents the number of initial tokens in arc (i, j).

Arcs have two valuations associated with them: $prod : E \rightarrow \mathbb{N}$ and $cons : E \rightarrow \mathbb{N}$. $prod(e)$ gives the constant number of tokens produced by $src(e)$ on e per firing and $cons(e)$ gives the constant number of tokens consumed by $snk(e)$ per firing.

A timed MRDF is defined by a tuple $(V, E, t, d, prod, cons)$.

Figure 3.1, we depict a MRDF graph with three actors A, B and C. Arrows represent arcs, with the arrowhead indicating the direction from production to consumption. Both arrow endpoints are annotated with the production/consumption rate of the producing/consuming actor in that arc. The black dots in an arc represent the delay (the number of initial tokens) on the arc.

Streaming applications often involve the processing of indefinitely long data sequences. Because of this, we are mainly interested in MRDF graphs that can be executed in a non-terminating fashion. Consequently, we must be able to obtain schedules that can run infinitely using a finite amount of physical memory. We say that a MRDF is *correctly constructed* if it can be scheduled periodically using a finite amount of memory. In this book, we will only concern ourselves correctly constructed MRDF graphs.

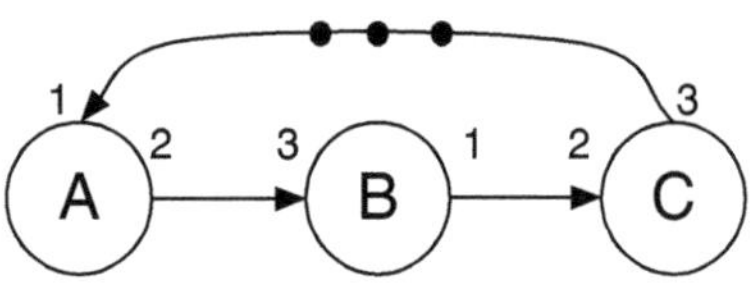

Fig. 3.1 Multi-rate data flow example

If there exists a fixed number of firings of each actor in a given MRDF graph that brings the graph to its initial token distribution and the graph is deadlock-free (how to check this will be explained in Sect. 3.3), then it can be scheduled periodically, and thus it can run within a finite amount of memory. If r is a vector of integer values such that each $r(i)$ represents the number of times that an actor i must be executed to bring the graph to its initial token distribution, then, for each edge $(i, j) \in E$ it must hold that the number of tokens produced over $r(i)$ firings of the producing actor i, and the number of tokens produced over $r(j)$ firings of the consuming actor j is the same, that is:

$$r(i).prod(i, j) = r(j) \cdot cons(i, j). \tag{3.1}$$

The set of equations thus obtained is normally referred to as the *balance equations* of a graph. It is convenient to represent these equations in matrix form.

The MRDF graph $G(V, E, t, d, prod, cons)$ can be represented by its *topology matrix*, T. The topology matrix contains one column per each node b in V and one row per each edge a in E. The value of element (a, b) is given by:

$$T(a,b) = \begin{cases} prod(e_a) & \text{if } src(e_a) = v_b \wedge snk(e_a) \neq v_b \\ -cons(e_a) & \text{if } snk(e_a) = v_b \wedge src(e_a) \neq v_b \\ prod(e_a) - cons(e_a) & \text{if } snk(e_a) = src(e_a) = v_b \\ 0 & \text{otherwise} \end{cases} \tag{3.2}$$

where $e_a \in E$ and $v_b \in V$, $a \in 1, 2 \ldots |E|$ and $b \in 1, \ldots |V|$.

The system of balance equations of Eq. 3.1 can then be expressed in matrix form as:

$$\boldsymbol{T}.\vec{r} = \vec{0} \tag{3.3}$$

where $\vec{0}$ is a column vector full of zeros, and $\vec{r}$ is a column vector.

The **repetition vector** for a correctly constructed MRDF graph with $|V|$ actors numbered 1 to $|V|$ is a column vector of length $|V|$, and corresponds to the smallest **positive integer** vector $\vec{r}$ which is a solution of Eq. 3.3. If each actor v_a is fired a number of times equal to the ath entry of $\vec{r}$, then the number of tokens per edge of the MRDF graph is equal to what it was in the initial state. The repetition vector **r** is useful for generating cyclic schedules for MRDF graphs. It will only exist if the MRDF graph is correctly constructed (see [56]). The repetition vector can be computed in polynomial time [56].

For the graph in Fig. 3.1 the repetition vector is $[3 \quad 2 \quad 1]$. If the production rate of actor C were 4 instead of 3, the balance equations would not have a solution other then $\vec{0}$, and the graph would not be correctly constructed.

An **iteration** of an MRDF graph is a set of actor firings such that each actor in the graph executes a number of times equal to its repetition vector entry. Therefore, in an iteration, each actor fires as many times as indicated by its entry in the repetition vector.

An execution of a MRDF graph is said to preserve the **FIFO** property if the order of production of tokens in output edges by different instances of an actor is the same as the order of consumption of these instances. This means that between any two firings of the same actor, the first one to consume inputs from the incoming edges is always the first one to produce outputs in the outgoing edges. Several firings can start at the same time (for instances, in the case of an actor that does not belong to any cycle, including a self-cycle, with enough input tokens in its input FIFOs to fire multiple times), but there will still be a firing for the first activation of the firing rules that consumes input tokens first and produces output tokens first, followed by a second firing that consumes second and produces second and so on. Thus, in a FIFO execution, the order of production of firings of an actor is deterministic and follows from the order of consumption. An MRDF is said to preserve the **First-In-First-Out (FIFO)** property if all its valid executions preserve the FIFO property.

Notice that it follows from the FIFO property that if $s(i,k)$ denotes the start time on a given execution of an MRDF graph of iteration k of actor i and $t(i,k)$ is the execution time of iteration k of actor i, then the MRDF is FIFO if for any execution of the MRDF graph it will hold that

$$s(i,k) + t(i,k) \leq s(i,k+1) + t(i,k+1). \tag{3.4}$$

guaranteeing that $s(i,k)$ can always produce its tokens in its output FIFOs before $s(i,k+1)$ (although perhaps at the same instant of time).

If an actor either has a constant execution time or belongs to a cycle with a single delay, any valid execution of the graph will conserve the FIFO property (see [3,78]). MRDF graphs under FIFO execution are functionally deterministic, since for each actor, independently of the time of arrival of tokens to the input queues, for the same values in the same order in the input queues, the same values are produced in the same order on the output queues. Because of this, in this book, we are only interested in executions of data flow graphs that preserve the FIFO property.

An MRDF graph is **monotonic** in time execution. Consider a FIFO MRDF graph $G = (V, E)$, with time valuation $t(i)$. Say that schedule function $s(i,k)$ represents the time at which the instance k of actor $i \in V$ is fired on a valid schedule of G. Consider now that $t(i)$ is replaced by a new valuation $t'(i)$, such that $t'(i) \leq t(i), \forall i \in V$. Monotonicity in time execution of G means that any valid schedule $s(i,k)$ for time valuation $t(i)$ is also admissible for time valuation $t'(i)$. Intuitively, earlier termination of an actor firing cannot cause other firings to occur later. This is because the firing rule of an MRDF actor is determined by the latest required input token to arrive and firing rules are fixed. If an input arrives earlier, it cannot cause the actor to fire later, and therefore it cannot cause the actor to produce outputs later.

3.3 Single Rate Data Flow

An MRDF graph in which, for every edge $e \in E$, it holds that $prod(e) = cons(e)$, is a **Single Rate Data Flow** (SRDF) graph. Any MRDF graph can be converted into an equivalent SRDF graph. Each actor i is replaced by $r(i)$ copies of itself, each

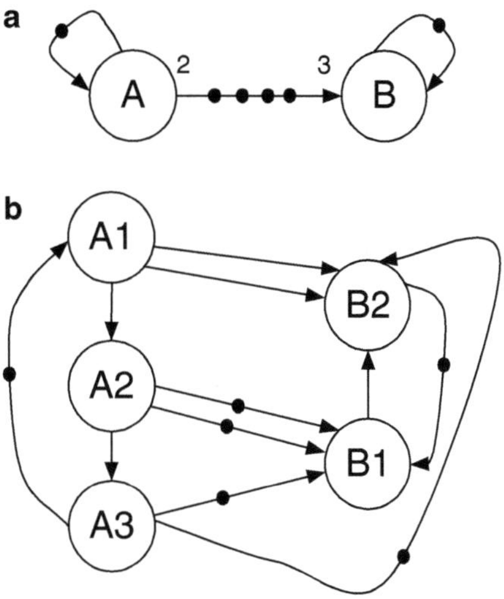

Fig. 3.2 An MRDF graph and its SRDF-equivalent graph

representing a particular firing of the actor within each iteration of the graph. That is, for any given actor i in the MRDF graph with an $r(i)$ entry in the repetition vector, if its copies in the equivalent SRDF graph are represented as i_p, $p = 0, 1 \ldots r(i) - 1$, the firing k of i_p corresponds to the firing $k.r(i) + p$ of the original MRDF actor i. The input and output ports of these nodes are connected in such a way that the tokens produced and consumed by every firing of each actor in the SRDF graph remains identical to that in the MRDF graph (see [81]).

Figure 3.2a depicts an MRDF graph. The SRDF-equivalent of this graph is depicted in Fig. 3.2b. The repetition vector of the MRDF graph is [3 2]. The SRDF-equivalent graph has as many copies of each actor as the number of repetitions of that actor in the repetitions vector. Each arc in the MRDF graph is represented by arcs between specific copies. The four initial tokens on the (A, B) edge are consumed by the first four consumptions of tokens by copies of B. This essentially determines which copies of A provide tokens for which copies of B.

The process of conversion will be explained in the next section in more detail, with respect to Cyclo-Static DF graphs, which are a superset of MRDF.

SRDF graphs have very useful analytical properties. A SRDF graph is **deadlock-free** if and only if there is at least one delay in every cycle [81]. A graph is **deadlocked** when there is a cyclic dependency where two actors cannot fire because each requires the other one to fire in order to obtain the data required for it to fire itself. The **cycle mean** of a cycle c in a SRDF graph is defined as $\mu_c = \frac{\sum_{i \in N(c)} t_i}{\sum_{e \in E(c)} d_e}$, where $N(c)$ is the set of all nodes traversed by cycle c and $E(c)$ is the set of all edges traversed by cycle c.

The **Maximum Cycle Mean (MCM)** $\mu(G)$ of a SRDF graph G is defined as:

$$\mu(G) = \max_{c \in C(G)} \frac{\sum_{i \in N(c)} t_i}{\sum_{(e \in E(c)} d_e} \tag{3.5}$$

where $C(G)$ is the set of simple cycles in graph G.

As we shall discuss in Chap. 4, the MCM of a SRDF graph is equivalent to the inverse of its maximum attainable throughput. Many algorithms of polynomial complexity have been proposed to find the MCM (see [17] for an overview).

3.4 Cyclo-Static Data Flow

Cyclo-static Data-Flow (CSDF) [9] can be seen as an extension of MRDF. In CSDF, each actor has a finite number of phases. Each firing of an actor corresponds to a phase. For each phase, the actor has a different production/consumption rate on each of its outgoing/incoming edges. The phase sequence occurs cyclically, always in the same order, and thus, for an actor with n phases, the production or consumption rates for each of the actor's outgoing or incoming edges at iteration k is the same as for iteration $(k \% n)$, where % represents the modulo operation (i.e. the remainder of integer division). Moreover, if the graph is timed, then each of its phases can have a different execution time.

Figure 3.3 depicts a CSDF graph. Actor C is a cyclo-static actor with two phases. This is indicated in the figure by representing the consumptions, productions and execution times of actor C as sequences of integer values. Thus, in its first phase, C consumes one token from edge (A,C), no tokens from edge (B,C), and produces one token in edge (C,D). In its second phase, it consumes one token each from edges (A,C) and (B,C), and produces two tokens on edge (C,D). All other actors behave as in a MRDF graph and are thus valid CSDF actors with a single phase.

In terms of notation, a timed CSDF graph can be represented by the tuple $(V, E, d, t, prod, cons)$. The set of timed CSDF graph actors is denoted by V and has a valuation $t : V \times \mathbb{N}_0 \rightarrow \mathbb{N}_0$, where, $t(i, k)$ is the execution time of actor i at iteration k. CSDF graph edges have a valuation $d : Ex \rightarrow \mathbb{N}$, where, $d(i, j)$ is the initial number of tokens (delays) of arc (i, j). Edges also have the following valuations: $prod : E \times \mathbb{N}_0 \rightarrow \mathbb{N}_0$ and $cons : E \times \mathbb{N}_0 \rightarrow \mathbb{N}_0$,

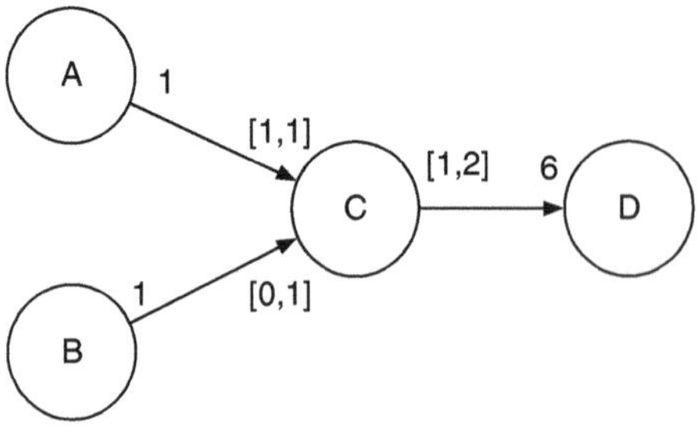

Fig. 3.3 Cyclo-static data flow example

where $prod((i,j),k)$ returns the number of tokens produced in iteration k on edge (i,j) and $cons((i,j),k)$ returns the number of tokens produced on edge (i,j) in iteration k. Furthermore, for each edge, it must be that the repeating sequence of production/consumption rates matches the number of phases of its producer/consumer. Thus, for every edge $(i,j) \in E$, there must exist two integers n_1 and n_2 such that $prod((i,j),k) = prod((i,j),k\%n_1)$, $t(i,k) = t(i,k\%n_1)$ and $cons((i,j),k) = cons((i,j),k\%n_2)$ and $t(j,k) = t(j,k\%n_2)$. The smallest non-zero value of n_1 and n_2 will be equivalent to, respectively, the number of phases of i and j. We will denote the number of phases of a given actor i as $nphases(i)$.

Similarly to MRDF graphs, CSDF graphs can only be consistent if there is a solution to the balance equations. For example, if in the graph depicted in Fig. 3.3 there would be an edge (C,A) with five delays, and $prod((C,A),0) = 2$, $prod((C,A),1) = 3$ and $cons((C,A),0) = 1$, the graph would never return to the initial state, as every two frings of C would enable five firings of A, thus resulting in five tokens to be produced in AC, while two firings of C can only consume two tokens in AC. No positive number of firings of A and C would ever return both edges to the initial delay distribution ($d(A,C) = 0$ and $d(C,A) = 5$) and the graph would not be consistent.

As the example illustrates, the balance equations for CSDF must take into account all phases of each actor. For each edge $(i,j) \in E$, the CSDF balance equation is:

$$q(i) \cdot \sum_{k=0}^{nphases(i)-1} prod((i,j),k) = q(j) \cdot \sum_{k=0}^{nphases(j)-1} cons((i,j),k). \tag{3.6}$$

where $q(i)$ represents the number of times the full phase sequence of i must be fired.

As an example, for the graph depicted in Fig. 3.3, a solution of the balance equation for edge AC (disregarding the other edges) would be $q(A) = 2$ and $q(C) = 1$, meaning that A must fire twice to produce enough data to be consumed by 1 full cycle of C phases, i.e. two firings of C.

A CSDF graph can also be denoted by a *topology matrix* T. In this topology matrix, the number of rows is equal to $|E|$ and number of columns is $|V|$. A number in the ith row and the jth column of T is equal to the difference between the sum of all productions and the sum of all consumptions of node j along edge i for the first $nrphases(i)$ iterations of i.

As for MRDF, the balance equations can be represented in matrix form by $T \cdot \vec{q} = \vec{0}$, where $\vec{q}$ is the vector solution of the matricial equation. If $\vec{q}$ is the smallest, non-zero solution to the matrix equation, it is referred to as the **phase repetition vector**. Note that there is always one trivial solution $\vec{q} = \vec{0}$. However, if apart from this trivial solution there is no solution, then the CSDF graph is inconsistent and cannot be scheduled in finite memory.

Given the phase repetition vector $\vec{q}$, then the repetitions vector for CSDF can be written as $\vec{r} = P \cdot \vec{q}$ where, $\vec{q}, \vec{r}$ are vectors of length $|V|$, and P is the diagonal matrix of the number of phases (*nphases(i)*) of each actor. An iteration of a CSDF graph occurs when each actor in the CSDF graph fires as many times as the number corresponding to it in the repetitions vector.

As for MRDF, if a CSDF graph is consistent, an SRDF equivalent can be constructed making use of the repetitions vector. As in the case of the MRDF to SRDF expansion algorithm, for each actor i, a number of SRDF actor copies i_n is created, one for each n such that $0 \leq n < r(i) \cdot nphases(i)$. Each actor copy i_n will be assigned an execution time such that $t(i_n) = t(i,n)$. Also, each actor copy i_n will have, for each incoming/outgoing edge in the CSDF graph, a number of incoming/outgoing edges equal to $prod(i,n)/cons(i,n)$. An edge $(i_{n\%q(i)}, j_{m\%q(j)})$ will be in the converted graph if (i, j) exists in the original CSDF graph, and iteration (i,n) of i produces a token that will be consumed by iteration (j,m) of j. Initial tokens in the CSDF edge (i, j) are distributed by the (i_n, j_n) edges in the converted graph such that the j_m copies corresponding to the first $d(i,k)$ consumptions in the edge consume these initial tokens.

The complete algorithm to perform this conversion is proposed in [9]. Often, however, it is useful to do the conversion by hand. A way of doing this is illustrated in Fig. 3.4 for two actors and one edge that are assumed to be part of a larger graph. The original CSDF actors and edge are depicted in Fig. 3.4a. Assume that after computing the repetitions we obtain that $q(A) = q(B) = 2$. In the first step of CSDF to SRDF conversion, for each actor, we must thus draw 4 copies of each actor, since $q(A) \cdot nphases(A) = q(B) \cdot nphases(B) = 4$, as depicted in Fig. 3.4b. In the same figure, we also draw stubs for the incoming and outgoing edges that in the SRDF graph will represent edge (A,B). For each copy A_k we draw a number of outgoing edge stubs equal to $prod((A, B), k)$. In a similar way, for each copy B_k we draw a number of incoming edge stubs equal to $cons((A, B), k)$. In the second step, depicted in Fig. 3.4c, we account for the consumption of initial tokens in the edge, i.e., in the original CSDF graph, B will first consume all initial tokens before consuming tokens produced by A. We thus start distributing the initial delays one per each input stub of B, starting from the copy with the lowest index to the copy with the highest index, and iterating as many times as necessary as to spend all initial tokens (i.e. if we place a token in the last stub of the last copy of the consumer, we go back to the first copy and continue distributing until we run out of initial tokens). In the final step, depicted in Fig. 3.4d, we connect the first stub of the first copy of A to the first input stub of a B copy after the last to receive an initial token. We then proceed to connect all stubs in order, iterating (i.e. going from last to first copy) if necessary, until we run out of outgoing edge stubs at the producer side.

3.5 Integer Data Flow

In his PhD thesis, Buck introduces Boolean Data Flow and Integer Data Flow. We will discuss Integer Data Flow only, as Boolean Data Flow can be seen as a special case of the former, where integer values are limited to 0 (false) and 1 (true). In Integer Data Flow (IDF), the ports of actors are labeled as being either data ports

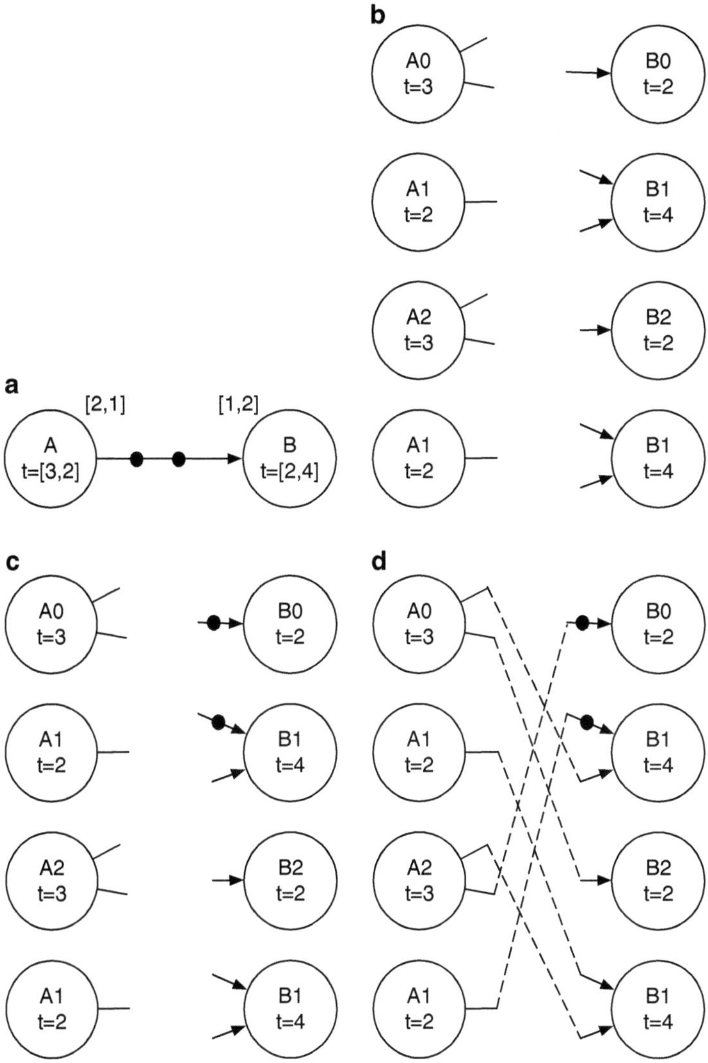

Fig. 3.4 A CSDF graph and its SRDF-equivalent graph

or control ports. If an actor has no control ports, it behaves just like a MRDF actor. If an actor has a control port, however, the consumption rates are no longer fixed. We shall refer to these actors as **variable-rate** actors.

Buck defines two types of control tokens:

- Type 1—case control: the number of tokens transferred by the variable-rate actor receiving this type of control token from/to each of its data in/out ports is either constant or zero, depending on the value of the control token.

- Type 2—repeat control: the number of tokens transferred by the variable-rate actor receiving this type of control token is determined by the value of the control token.

Although many variable rate actor behaviors are possible, we will only describe the canonical constructs employed by Miyazaki and Lee [69]: Switch (Case), Select(End-Case), Repeat Begin and Repeat End. These are depicted in Fig. 3.5. The control port is marked with a triangle. Any IDF behavior can be represented employing only these canonical variable-rate actors and MRDF actors. The behavior of the canonical variable rate actors is defined as follows:

- **Switch** has one input control port that reads type 1 control tokens, one data input port with a fixed rate of one and a number of variable-rate output ports. Each output port is **keyed** to a specific integer value, with the exception of one output port, the "default" output port. The actor is fired whenever one input token is available in both input ports. When the actor fires, it consumes both input tokens and copies the value of the data input token to the output port keyed to the value of the control token. If there is no port keyed to the value of the control token, then the input data token is produced in the "default" output port.
- **Select** has one input control port that reads type 1 control tokens, it has a number of variable-rate input ports, and one output port with a fixed rate of one. Each input port is keyed to a specific integer value, with the exception of one input port, the "default" input port. The actor is fired whenever one input token is on the control port, and one input token is available in the input port keyed to the value of the input token on the control port. If no input port exists keyed to the value of the input token on the control port, then the actor requires one token on the "default" input port. When the actor fires, it consumes both input tokens and copies the value of the data input token to the output port.
- **Repeat Begin** has one input control port that reads type 2 control tokens, one data input port with a fixed rate of one, and one data input port with a variable rate. Repeat Begin fires when a token is available in each of its input ports. It consumes both input tokens and produces as many copies of the data input token as the integer value of control input port.
- **Repeat End** has one input control port that reads type 2 control tokens, one data input port with variable rate, and one output port with a fixed rate of one. Repeat End fires when a token is available in the control input port and N tokens are available in the data input port, where N is the integer value of the token in the control input port. When it fires, it consumes all the input tokens required for the firing and produces one single token in the output port.

IDF graphs are still deterministic: the series of values produced in all its FIFOs is always the same, given the same series of input values. Also, given a specific series of input values, the behavior of IDF is monotonic. The firing time of an actor instance is dependent on the maximum of the arrival times of the input tokens. Since IDF is deterministic, the time at which a token arrives cannot change the values of tokens, including control tokens. Therefore, all actor firings have the same firing

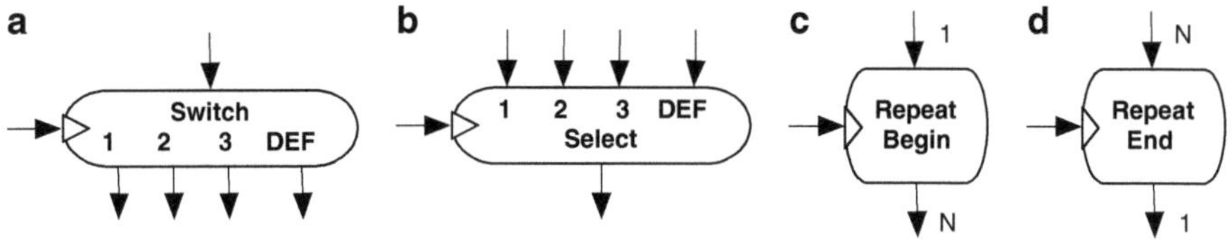

Fig. 3.5 Four variable-rate actors used in IDF graphs

rules independently of the time at which each firing happens. Therefore, the earlier arrival of a token can still only cause its consuming actor to fire earlier, and IDF graphs are monotonic.

Note that actors whose consumption rates depend on the value of the input control token, such as Select and Repeat End, must be evaluated in two phases, since the firing rules can only be determined by reading the input control token.

In terms of expressivity, Integer DF is Turing-complete. This means that not all IDF graphs can be statically scheduled, and for many IDF graphs it is impossible to say whether they can execute within bounded memory, or whether they deadlock or not, because cyclic schedules that are unbounded in length, or that may require unbounded memory on arcs, may occur or not, depending on the values of control tokens.

Buck [10] shows how to solve the balance equations of an IDF graph to determine whether bounded-length schedules exist. We will briefly show how this is handled since this concept will be necessary to prove the properties of our own model, Mode-Controlled Data Flow, in Chap. 6.

Buck solves the balance equations for dynamic data flow graphs by using symbolic expressions for the number of tokens produced and consumed on conditional arcs. For IDF, these symbolic expressions are of the form p_{ij}, and represent the proportion of integer control tokens on control stream c_i whose value is j.

The repetition vector can then be obtained by solving symbolically the balance equations. This will result in a repetition vector $\vec{r}$ that is a function of all the p_{ij} variables. If there are nonzero solutions for the repetition vector regardless of the value of all the p_{ij}, then the graph is said to be **strongly consistent**. A strongly consistent IDF graph has a balance of long-term flow rates. If we interpret the p_{ij} expressions as long-term average rates, strong consistency asserts that the rates are in balance regardless of the precise proportions of the control tokens that assume each integer value. This, however, says nothing about the existence of bounded-length schedules or bounded-memory implementations. In [10], examples of strongly consistent graphs that require unbounded memory are given.

Buck also introduces the concept of **bounded complete cycle**. A complete cycle of an IDF graph is a sequence of actor firings that returns the graph to its original state. A null sequence is a trivial solution. A **minimal complete cycle** is a non-null complete cycle without a non-empty sub-sequence that is also a complete cycle.

To determine the properties of complete cycles, one must solve the balance equations. Since $T.\vec{r} = \vec{0}$, if we fire actors such that each actor i is fired r_i times,

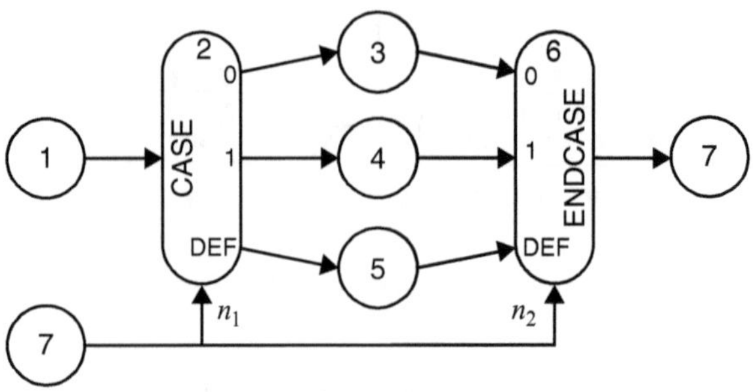

Fig. 3.6 Integer data flow example (Taken from [10])

the system returns to the original state. If there is only a trivial solution to the balance equations (i.e. 0 firings for all actors) then we conclude that no minimal complete cycles exist. If the balance equations have nontrivial solutions, then either the graph deadlocks, or schedules that are bounded both in length and in memory requirements exist.

If the p_{ij} expressions are interpreted as the fraction of control tokens on control stream c_i during a complete cycle that have value j, then $p_{ij} = \frac{t_{ij}}{n_i}$, where n_i is the total number of control tokens produced on a complete cycle, and t_{ij} is the total number of these that have value j. The properties of complete cycles can then be analyzed by symbolically solving the balance equations as explained before, then replacing p_{ij} by $\frac{t_{ij}}{n_i}$ and then constraining the number of actor firings, control tokens, and all t_{ij} variables to be integral. Minimal complete cycles must satisfy the balance equations.

Even if bounded solutions exist for the balance equations, it may be that there is no schedule for the thus obtained repetitions vector that is deadlock-free. Verifying deadlock freedom of the minimal complete cycles is done by checking if precedence constraints cannot prevent the firing of each actor for the number of times required by the minimal complete cycle. This is done by specifying the exact values of the emitted control tokens, for every control sequence that realizes the minimal complete cycle. The IDF graph is therefore transformed onto a set of MRDF graphs, one for each control sequence, and for each deadlock-freedom can be independently verified.

If an IDF graph has even a single arc of type 2, it immediately has unbounded memory, because there is no limit on how large an integer control token can be. Bounds can only be given by knowing the maximum value of the control tokens.

Consider the example in Fig. 3.6 (taken from [10]). All actors are SRDF except for the CASE (Switch) and ENDCASE (Select).

By solving symbolically the balance equations as explained, we obtain the following expression for the repetitions vector:

$$\vec{r}(\vec{p}) = k[1 \quad 1 \quad p_{10} \quad p_{11} \quad (1 - p_{10} - p_{11}) \quad 1 \quad 1 \quad 1]^T. \qquad (3.7)$$

As explained, p_{10} can be interpreted as the number of tokens on control stream 1 (i.e. the tokens produced on edge 8) with value 0, during a complete cycle divided by the number of tokens n_1 on stream 1 in a complete cycle. We can then find the smallest integer solution. We find out that there is only one control token per complete cycle and the repetition vector is

$$\vec{r}(\vec{p}) = [1 \quad 1 \quad t_0 \quad t_1 \quad (1 - t_0 - t_1) \quad 1 \quad 1 \quad 1]^T, \tag{3.8}$$

where t_0 is 1 if the control token is 0 and 0 otherwise, and t_1 is 1 if the control token is 1 and 0 otherwise.

There are three complete cycles. One where control sequence $c_1 = 0$, another where $c_1 = 1$ and another where $c_1 = X$, where X is any integer value other than 0 or 1.

It is easy to verify that all complete cycles are deadlock-free, since the graph has no cycles. Please note that if the control streams on n_1 and n_2 were independent, then there would not be any bounded complete cycles, and the graph would not be strongly consistent.

The considerations we made for IDF also hold for Boolean DF, which can be seen as a special case of IDF where the control tokens are limited to having values of 1 or 0. The main difference is that in Boolean DF there are no type 2 control edges. Besides that, all properties are the same.

To summarize, in order to analyze the execution behavior of IDF graphs, one must first symbolically compute the repetitions vector. If the graph is strongly consistent, one can proceed to find minimal bounded cycles. If these exist, the graph can execute in bounded memory. Therefore, for some graphs, deadlock-freedom, and bounded memory execution can be proved to exist. For others, it is not even possible to say whether they deadlock or not, as this is fully dependent on the values of the control tokens produced in each control stream.

3.6 Conclusion

When choosing a data flow model as a programming or analysis model, one must make a trade-off between expressivity and analytical properties. An essential difference between data flow flavors is whether the chosen model allows for data dependent behavior or not. If all production and consumption rates of all firings of all actors can be determined independently of the values contained on the input streams, then a Static Data Flow variant will suffice. These include the SRDF, but also the MRDF and CSDF models, which can also be converted to a SRDF-equivalent graph. These models allow automatic checking of deadlock freedom in polynomial time on the size of the SRDF-equivalent graph, static scheduling and timing analysis, if actors are time-annotated. If the firing rules of actors are dependent on the values of the input data, then we need a flavor of Dynamic Data Flow. Models such as IDF and BDF are Turing-complete, meaning that it may not be possible to even check whether a specific graph can deadlock or not, let alone perform timing analysis. The model we will describe in Chap. 6 tries to retain the analytical properties of static data flow while allowing for data dependent behavior.

Chapter 4
Temporal Analysis

In Chap. 2, we concluded that, due to the application requirements we are aiming at, we cannot schedule jobs statically. There are several reasons for this: first, actors have variable execution times, second, the execution of the graph is performed across the multiprocessor, and it may be difficult to guarantee that all processors are synchronized at all times, and also, we cannot statically schedule together two independent transceiver jobs. Therefore, synchronization in our implementation is, at the essence, self-timed: a data flow actor fires immediately whenever its firing conditions are met.

Temporal analysis is required in order to verify whether a given timed data flow graph can meet a throughput or latency requirement of the application. In Chap. 5 we will show how we are able to model mapping decisions on a data flow graph that represents the temporal behavior of a given mapping to a platform of an application graph. This will allow for a mapping flow where every mapping decision can be translated onto a transformation of the application graph, and evaluated with respect to its temporal behavior, using data flow analysis techniques. In other words, we derive from the functional graph of the transceiver an implementation-aware graph where we model worst-case assumptions about the timing of actor firings and communication in a given platform. Since the execution model of our platforms of interest (see Chaps. 1 and 2) is self-timed (i.e., execution of actors is triggered by arrival of data and availability of output space), we are mostly interested in analyzing the temporal behavior of the self-timed execution of data flow graphs. Furthermore, as it will be shown, many common scheduling approaches can be modeled in the self-timed execution model, by imposing additional constraints to actor execution.

In this chapter, we propose techniques for the analysis of the self-timed behavior of the data flow variants with static rates. Our techniques operate directly on Single-Rate Data Flow graphs, and can easily be adapted to any other static data flow variant that can be converted to Single-Rate Data Flow, such as Cyclo-Static Data Flow, or Multi-Rate Data Flow.

Previous techniques [81] for the characterization of the temporal behavior of the self-timed execution of SRDF graphs were geared towards giving guarantees with

O. Moreira and H. Corporaal, *Scheduling Real-Time Streaming Applications onto an Embedded Multiprocessor*, Embedded Systems 24, DOI 10.1007/978-3-319-01246-9_4,

respect to guaranteed throughput over a long run of iterations of the graph. This is not enough for real time streaming applications. For example, if the application is a radio transmitter, we may need to guarantee that we can generate output at a strictly periodic rate, from the first output on, without experiencing any hiccups at the output. In some cases, such as for a TD-SCDMA or WLAN receiver, we need to guarantee a certain maximum latency between the reception of a message and the production of an acknowledgment. Previously proposed analysis of self-timed schedules can only tell us about average throughput, and only after the execution has converged into periodic behavior. Although it has been proven that the self-timed execution of a data flow graph will eventually reach a regime with periodic behavior [23], determining how many actor firings are required for this to happen is not efficiently computable [3]. Moreover, periodic behavior can only be attained if the actors have constant execution times.

The problem becomes more complex because a receiver job may be driven by an external source that does not exhibit a periodic behavior. For instance, in WLAN 802.11a, packets arrive sporadically, i.e., we know the minimum time interval between the arrival of packets, but not the maximum time interval.

In this chapter we will present techniques that allow us to reason about the temporal behavior of the self-timed execution of static data flow graphs, even when considering the transient phase and/or varying execution times per actor firing. We will define sufficient conditions to achieve a strictly periodic behavior for an output, while respecting the behavior of an external source that forces its own strictly periodic behavior. We will provide a definition of latency and techniques to characterize maximum latencies in the self-timed execution of the data flow graph, even in the presence of aperiodic external sources. To do this, we will establish an important relation between self-timed execution and static periodic schedules, that allows us to use the generation of static periodic schedules as a means to bound the behavior of a self-timed execution. We will enunciate and prove the existence of an important property of the self-timed execution of data flow graphs, the linear timing property. A form of this property is essential to handle external sources with sporadic behavior.

4.1 External Sources in Data Flow

In our data flow graphs, we will model external sources and sinks as actors. External sources and sinks are different from other actors in that the time at which the firing of such actors occurs may be imposed by the environment. One cannot delay the firing of an external source actor. It is important that we have this in mind when we talk about analysis and scheduling. A schedule that does not allow an external periodic source to execute in a strictly periodic regime from the first firing on is not valid. In fact, the external source actor firing pattern is defined independently of the schedule of the data flow, and either the schedule accommodates for the behavior of the source or it will be invalid. We can also think of it as the timing behavior of the external source effectively imposing a timing constraint on the execution of the graph.

Another problem with external sources is that they may not have a worst-case execution time – such is the case for a sporadic source. It must be also noted that the function of a source actor cannot be assumed to be functionally deterministic, as the stream of tokens produced by the source can be different from one execution to another.

Moreover, although the firing of the source cannot be controlled, its correct operation may still be dependent on values produced by the data flow graph. This is the case, for instance, when the source of the system is the output of an amplifier and the gain of that amplifier needs to be adapted by feedback from the function being computed by the digital transceiver. Still, the firing of the actor that represents the amplifier cannot be delayed, and for correct operation, the new gain must be available in time. This means that although the firing of the source is in practice independent from any internal event, there can be edges in the data flow graph from another actor to the actor that represents a source. It is part of the temporal requirements of the application that the data dependencies described by such edges are met in such a way that guarantees that the source has the correct input tokens ready at its forced start time. In this chapter, we will assume the source to be represented by a single actor with a self-edge containing one delay. If the execution time of this actor is constant, the source will be periodic with a period equal to each execution time. In Chap. 5, we will introduce two examples of WLAN and TD-SCDMA receiver that have sources with cyclo-static behavior, i.e., the source has several phases, each potentially taking a different, constant time to fire and each producing tokens to a different output edge. We will model such sources as a number of actors equal to the number of phases, connected by edges in a single cycle, with a token in the edge preceding the source actor which should fire first. Such sources can be analysed by the techniques proposed in this chapter, simply looking at the first actor in the source cycle as the single strictly periodic external source.

4.2 Schedules

As discussed in Sect. 3.2, MRDF graphs where actors have constant execution times can be statically scheduled because the firing conditions of each actor are independent of the values of the data tokens consumed, and, for well-constructed graphs that are deadlock-free, it is possible to find a number of firings per actor that allows the graph to return to its initial state, thus guaranteeing that there is a static schedule that can be executed continuously within bounded memory.

However, as discussed in Chap. 2, the fact that we must cope with variation of the execution times of actors, the effects of synchronization across a multi-processor, and of local inter-job schedulers, we cannot rely on static scheduling. Our proposed analysis strategy is based on representing the worst-case behavior of all implementation details of data flow, and then performing analysis of the self-timed execution of such a graph. In self-timed execution, each actor fires immediately whenever possible. If all timing constraints have been conservatively expressed in the data flow model, the self-timed execution of the graph will reflect the worst case temporal behavior of the implementation. In order to bound the start times of firings on a self-timed execution, we will determine in this section a relation between self-timed execution and static periodic schedules. The reader should be aware that static periodic schedules are not used here as a form of implementation, but as a technique to provide useful bounds on self-timed execution.

Before we go into that, however, we must introduce a scheduling notation, and define both self-timed and static periodic schedules, and their properties.

4.2.1 Notation

The schedule function $s(i,k)$ represents the time (we will use positive real numbers to represent time) at which the instance k of actor i is fired. The instance number is counted from 0 and, because of that, the instance k corresponds to the $(k+1)$th firing. Furthermore we denote the finishing time of firing k of actor i by $f(i,k)$ and the execution time of firing k of i by $t(i,k)$. It always holds that $f(i,k) = s(i,k) + t(i,k)$. If $t(i)$ is the WCET of actor i, then $t(i,k) \le t(i), \forall k \in \mathbb{N}_0$.

4.2.2 Admissible Schedules

A schedule is admissible if, for every actor in the graph, and every firing of the actor, start times do not violate the firing rules. In [32] a theorem is given that states a set of necessary and sufficient conditions for an admissible schedule, assuming constant execution times:

Theorem 4.1. *A schedule s is admissible if and only if for any arc (i,j) in the graph and any $k \in \mathbb{N}_0$:*

$$s(j,k) \ge s\left(i, \left\lceil \frac{(k+1)\cdot cons(i,j) - d(i,j) - prod(i,j)}{prod(i,j)} \right\rceil\right) + t(i). \quad (4.1)$$

When applied to an SRDF graph, this inequality becomes simply

$$s(j,k) \ge s(i, k - d(i,j)) + t(i), \quad (4.2)$$

as productions and consumptions are 1 in all arcs. Since the schedule function is not defined for iterations lower than 0, an arc (i,j) only results in a constraint on the start time of j if $k \ge d(i,j)$.

We also assume that time starts at $t = 0$ and therefore, for a valid schedule, we will require that for any $i \in V$ and for any iteration $k \in \mathbb{N}_0$, the start time $s(i,k) \geq 0$. Moreover, we will make the assumption that time is continuous, i.e. $s(i,k) \in \mathbb{R}$.

For an MRDF graph converted into an SRDF-equivalent graph for analysis purposes, a relation between the start times of the SRDF copies of an original MRDF actor can be established easily. Say that i_p is the copy number p of an MRDF actor i in the equivalent SRDF graph. Then $s(i_p,k) = s(i, k \cdot r(i) + p)$. The same relation holds for CSDF graphs.

From here on, scheduling will be discussed, for the sake of simplicity, on SRDF graphs. We will, when appropriate, explain how concepts translate to MRDF or CSDF graphs.

4.2.3 Self-Timed Schedules

A **Self-Timed Schedule** (STS) of an SRDF graph is a schedule where each actor firing starts immediately when there are enough tokens on all its input edges.

The **Worst-Case Self-Timed Schedule (WCSTS)** of an SRDF is the self-timed schedule of an SRDF where every iteration of every actor i takes $t(i)$ time to execute. Note that the WCSTS of an SRDF graph is unique.

In an STS, the start times of actors occur as soon as all precedence constraints (as given by Eq. 4.2) are met, that is, for any given actor $i \in V$, the start time of firing k in the worst-case self-timed schedule s is

$$s(i,k) = \max_{(x,i)\in E} \begin{cases} s(x, k - d(x,i)) + t(x, k - d(x,i)), & k \geq d(x,i) \\ 0, & k < d(x,i) \end{cases} \tag{4.3}$$

For the WCSTS, $t(x, k - d(x,i))$ is a constant equal to the worst-case execution time of x, denoted by $t(x)$.

The WCSTS of a strongly-connected SRDF graph has an interesting property: after a transition phase of $K(G)$ iterations, it will reach a periodic regime. K(G) is a constant for a given timed SRDF. It can be determined by simulating the execution of the timed SRDF and detecting the start of the periodic behavior. This periodic regime has a period of $N(G).\mu(G)$ time units, where $N(G)$ is the **cyclicity** of the SRDF graph, as defined in [3]. To compute $N(G)$, find all cycles of G that have a cycle mean equal to the MCM; derive a graph G_c from G, by taking all nodes and edges of G traversed by critical cycles; compute the strongly connected components of G_c; for each strongly connected subgraph G_s of G_c, compute the sum of tokens d_p for every simple path p. Let d_s be equal to the gcd of all d_p in a given G_S. $N(G)$ is equal to the lcm of the d_s values for all the G_S components of G_c.

The schedule for the periodic regime is:

$$s(i, k + N(G)) = s(i,k) + N(G) \cdot \mu(G), \forall k \geq K(G). \tag{4.4}$$

During periodic execution, $N(G)$ firings of i happen in $N(G) \cdot \mu(G)$ time, yielding an average throughput [3, 23] of $1/\mu(G)$. For the transition phase, that is, for $k < K(G)$, the schedule can be derived by simulating the execution of the data flow graph, given WCETs of all actors. Another known method for calculating K(G) is presented in [3].

4.2.4 Static Periodic Schedules

A **Static Periodic Schedule** (SPS) of an SRDF graph is a schedule such that, for all nodes $i \in V$, and all $k > 0$

$$s(i,k) = s(i,0) + T \cdot k, \tag{4.5}$$

where T is the desired period of the SPS. Please note that an SPS can be represented uniquely by T and the values of $s(i,0), \forall i \in V$.

Theorem 4.2. *For an SRDF graph $G = (V, E, t, d)$, it is possible to find an SPS schedule, if and only if $T \geq \mu(G)$. If $T < \mu(G)$, then no SPS schedule exists with period T.*

Proof. Recall that according to Eq. 4.2 we know that every edge (i, j) in the data flow graph imposes a precedence constraint of the form $s(j, k+d(i, j)) \geq s(i,k)+t(i)$ to any admissible schedule. Since the start times in an SPS schedule are given by 4.5 we can write for every edge $(i, j) \in E$ a constraint in the form:

$$\begin{aligned} s(j,0) + T \cdot (k + d(i,j)) &\geq s(i,0) + T \cdot k + t(i) \\ \Leftrightarrow s(i,0) - s(j,0) &\leq T \cdot d(i,j) - t(i). \end{aligned} \tag{4.6}$$

These inequalities define a system of difference constraints – a special case of linear constraints where all constraints are constant differences between two variables. According to [15] this system has a solution if and only if the constraint graph does not contain any negative cycles for weights $w(i, j) = T.d(i, j) - t(i)$.

Since the MCM $\mu(G)$ is defined as (see Sect. 3.3):

$$\mu(G) = \max_{\forall c \in C(G)} \frac{\sum_c t(i)}{\sum_c d(i,j)} \tag{4.7}$$

then, for each cycle $c \in C(G)$ it must hold that:

$$T \geq \frac{\sum_c t(i)}{\sum_c d(i,j)}. \tag{4.8}$$

The inequality 4.8 can be rewritten as

$$\sum_c (T \cdot d(i,j) - t(i) \geq 0, \tag{4.9}$$

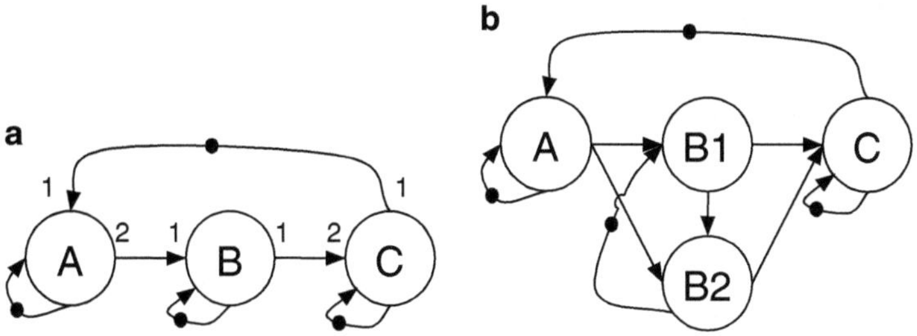

Fig. 4.1 An MRDF graph and its SRDF-equivalent graph

that is, if $T \geq \mu(G)$, there are no negative cycles for weights $w(i, j) = T \cdot d(i, j) - t(i)$ and, therefore, the system given by 4.6 has solutions. □

Therefore, $1/\mu(G)$ is the fastest possible rate (or throughput) of any actor in the strongly-connected SRDF. For an actor a of a CSDF graph G, it means that each one of its copies a_i in the SRDF-equivalent graph G' can execute at most once per $\mu(G')$. Therefore, the average rate of a per graph iteration is upper-bounded by $r(a) \cdot 1/\mu(G')$, although the execution is not necessarily strictly periodic (see below).

If a SPS has a period T equal to the MCM of the SRDF graph $\mu(G)$, we say that the schedule is a **Rate-Optimal Static Periodic Schedule (ROSPS)**. An SPS for a given G and T can be found by solving the system of linear constraints given by Eq. 4.6.

A solution can be found for any given $T \geq \mu(G)$ by using a single-source shortest-path algorithm that can cope with negative weights, such as Bellman-Ford [81], but other solutions may exist for any given graph and period.

Notice that, for an MRDF or CSDF graph, the SPS schedule of its SRDF-equivalent graph specifies an independent periodic regime for each SRDF copy of an original CSDF actor, but it enforces no periodicity between firings of different copies. If a strictly periodic regime with period $T/r(a)$ is required for actor a, extra linear constraints must be added to the problem. In some cases, this will result in an infeasible problem.

Consider the MRDF graph in Fig. 4.1a, and assume that all actor firings take 1 unit of time. The repetition vector for this graph is $[1 \quad 2 \quad 1]$. The SRDF-equivalent of the graph is depicted in Fig. 4.1b. The MCM of this graph is 4, given by the cycle $c = ((A, B1), (B1, B2), (B2, C), (C, A))$. If we try to impose a static period of $4/2$ between the firings of B, one must require that $s(B1, k) = s(B2, k) + 2$. Such schedule does not exist because forcing a distance of 2 time units between the two firings of B will always force cycle c to take at least 5 units of time to execute, thus effectively making it impossible to find a SPS with period equal to the MCM of the SRDF-equivalent graph.

4.2.5 Monotonicity

We have already seen that it is possible to construct an SPS of any SRDF graph with a throughput equal to $1/\mu(G)$ and that the WCSTS will eventually settle into periodic behavior with an average throughput equal to $1/\mu(G)$. Calculating $\mu(G)$ or trying to find an SPS schedule with period $\mu(G)$ are two ways to check for desired throughput feasibility. Two essential questions are yet to be answered: what happens during the transition phase, and how does STS behave with variable execution times? One property of SRDF graphs that allows us to give answers to these questions is monotonicity.

Because of the monotonicity of self-timed execution, if any given firing of an actor finishes execution faster than its worst-case execution time (WCET), then any subsequent firings in any self-timed schedule can never happen later than in the WCSTS, which can be seen as a function that bounds all start times for any self-timed execution of the graph. We can enunciate this as a theorem:

Theorem 4.3 (Monotonicity of self-timed execution). *For an SRDF graph $G = (V, E, t, d)$ with worst-case self-timed schedule s_{WCSTS}, for any $i \in V$, and $k \geq 0$, it holds that, for any self-timed schedule s_{STS} of G*

$$s_{STS}(i,k) \leq s_{WCSTS}(i,k). \tag{4.10}$$

Proof. The start times of all firings of actor i in any self-timed schedule are given by Eq. 4.3. Since for all edges $(x,i) \in V$ it is true, by definition, that $t(x, k-d(x,i)) \leq t(x)$, then it is true that

$$s_{STS}(i,k) \leq \max_{(x,i)\in E} \begin{cases} s_{STS}(x, k-d(x,i)) + t(x), & k \geq d(x,i) \\ 0, & k < d(x,i). \end{cases} \tag{4.11}$$

If the theorem holds for the predecessors of firing k of i, we may write $s(x, k - d(x,i)) \leq s_{WCSTS}(x, k-d(x,i))$. Then it is true that

$$s_{STS}(i,k) \leq \max_{(x,i)\in E} \begin{cases} s_{WCSTS}(x, k-d(x,i)) + t(x), & k \geq d(x,i) \\ 0, & k < d(x,i) \end{cases} \tag{4.12}$$

and this inequality is equivalent to

$$s_{STS}(i,k) \leq s_{WCSTS}(i,k). \tag{4.13}$$

Therefore if the theorem is valid for all the firings on which firing k of i depends, it is also valid for firing k of i. As for the initial elements, the first actors to execute are the ones that do not have initial edges, or only have initial edges with initial tokens. For those, the start times are 0 for both any s_{STS} and for the s_{WCSTS}. □

4.2.6 Relation Between the WCSTS and SPS

Because of monotonicity, assuming worst-case execution times, the start time of any actor cannot happen earlier than in the WCSTS: since in a STS firings happen as early as possible, there is no way to schedule anything earlier without violating the firing rule. In other words, any valid schedule of an SRDF graph that assumes worst case execution times for all firings must have its start times later or equal to the start times given by the WCSTS. Since valid Static Periodic Schedules must assume worst-case execution times, the following theorem holds:

Theorem 4.4. *In any admissible SPS of an SRDF graph $G = (V, E, t, d)$, all start times can only be later or at the same time than in the WCSTS of that graph, that is, for all $i \in V$, $k \geq 0$, and all admissible static periodic schedules s_{SPS} of G, it must hold that*

$$s_{WCSTS}(i,k) \leq s_{SPS}(i,k). \tag{4.14}$$

From this we draw an important conclusion: for a given SRDF graph, **any SPS start time can be used as an upper bound to any start time of the same firing of the same actor in the WCSTS and, due to Theorem 4.3 in any STS of the graph**).

This means that, given an SRDF graph for which we know the worst-case execution times of all actors, we can obtain a linear conservative upper bound on the output production times of any given actor i during self-timed execution. This bound is given by the expression $s(i,0) + t(i) + \mu_G.k$, where $s(i,0)$ is the start time of the first firing of i on a Rate-Optimal Static Periodic Schedule (ROSPS).

To find the lowest bound, one can compute a ROSPS that minimizes $s(i,0)$. A single schedule can jointly minimize $s(i,0)$ for all actors because, by inspection of the sequence constraints given by Eq. 4.6, one can see that reducing the initial start time of an actor can never constrain the start time of another actor with a higher lower bound. We can formulate an LP program that accomplishes the lowest linear bounds and minimizes the sum of start times of first firings of actors, subject to the sequence constraints in Eq. 4.6 and positive start times:

Let $G = (V, E, t, d)$.	
Minimize	$\sum_{i \in V} s(i,0)$
subject to	
$\forall (i,j) \in E$,	$s(j,0) - s(i,0) \geq t(i) - \mu_G \cdot d(i,j)$,
$\forall i \in V$,	$s(i,0) \geq 0$.

Note that we can also use an LP to find an SPS that minimizes only one particular start time (in a LP, the optimization function is a sum of variables multiplied by real coefficients – we can make all but one equal to 0). Doing this for each start time will provide us with the tightest linear bound for the firings of each actor in the graph.

4.3 Linear Timing of Self-Timed Execution

We will now introduce the Linear timing property. This property will allow us to infer how forcing an actor to fire later than its self-timed firing time will affect the timing of other firings.

Definition 4.1 (Linear timing). A data flow graph $G = (V, G, d, t)$ is linearly-timed if, given a schedule $s(i,k)$ where all firings are self-timed, and actor execution times are given by $t(i,k)$, then, if a new schedule s' is constructed where all firings are self-timed, except for the firing p of actor $n \in V$, which is given by $s'(n, p) = s(n, p) + \delta$ then, for all $i \in V$ and all $k \in \mathbb{N}_0$, $s'(i,k) \leq s(i,k) + \delta$.

Theorem 4.5. *A well-constructed, deadlock-free SRDF graph G=(V,E,t,d) is linearly-timed.*

Proof. Consider an STS schedule s for G, with execution times given by $t(i,k) \leq t(i)$. Then, for all actors $i \in V$, and for all $k \in N_0$, the start times are given by Eq. 4.3.

For the schedule s' in Definition 4.1, the start times are given by

$$s'(i,k) = \max_{(x,i)\in E} (s'(x,k-d(x,i)) + t(x,k-d(x,i))), \tag{4.15}$$

with the exception of $s'(n,p) = s(n,p) + \delta$.

We can now prove by induction: if any number of the direct predecessors of firing k of i is delayed by at most δ, we may write that $s'(x,k-d(x,i)) \leq s(x,k-d(x,i)) + \delta$, for all predecessors. It is then true for any i that

$$s'(i,k) \leq max_{(x,i)\in E}(s(x,k-d(x,i)) + t(x,k-d(x,i)) + \delta). \tag{4.16}$$

Since δ is added to all arguments of the max expression, we may pull it outside of the max expression and rewrite the inequality as

$$s'(i,k) \leq max_{(x,i)\in E}(s(x,k-d(x,i) + t(x,k-d(x,i))) + \delta \tag{4.17}$$

and due to Eq. 4.3 we may replace the max expression in Eq. 4.17 by $s(i,k)$ to obtain

$$s'(i,k) \leq s(i,k) + \delta. \tag{4.18}$$

□

The linear timing property allows us to establish relations between self-timed schedules: the effect on all other firings of delaying the (n,p) firing by δ is equivalent to increasing the execution time $t(n,p)$ by δ. It can also be easily proven that a negative δ can maximally decrease the start times of all actors by δ.

4.4 Dependence Distance

Not all firings can be affected by a change in the start time of firing (n, p). In fact, only firings that are dependent on tokens produced by firing p of actor n can be affected by the increment. We will capture this notion in the concept of dependence distance. Intuitively, the dependence distance $dd(i,k)$ tells us which is the highest iteration number k' of actor j that can fire before iteration $k+1$ of i fires, thus $dd(i,j) = k' - k$.

Definition 4.2. The **Dependence Distance** $dd(i, j)$ represents the number of firings of actor $j \in V$ that can occur before the first firing of actor $i \in V$, in an admissible schedule. That is, for any admissible schedule, $s(j, dd(i, j)) \geq s(i, 0) + t(i, 0)$. Because SRDF actors have unary rates of production and consumption, it also means that $s(j, k + dd(i, j)) \geq s(i, k) + t(i, k)$.

How can we compute the dependence distance between two actors? Intuitively, in a strongly connected graph, if we try to execute all firings of all actors in the graph in a self-timed manner, except for i, which we do not allow to fire, then we can simply count the number of firings that were possible for each other actor before the graph deadlocks, thus obtaining $dd(i, j)$ for all $j \in V - \{i\}$.

If there is no path from i to j (and the graph is therefore not strongly connected) the dependence distance is infinite (meaning that an infinite amount of firings of j can occur before any firing of i). This, however, cannot occur in strongly connected graphs.

If i never fires, the maximum amount of tokens that can reach an input arc of j that belongs to a direct path from i is equivalent to the amount of tokens in that path. If there are many such paths, then j can only fire a number of times equal to the minimum amount of initial tokens in any of those paths.

The value of $dd(a, c)$ is therefore equivalent to the shortest-path in the graph for weights equal to the $delay$ of edges. Since delay is always non-negative, the shortest paths can only be non-negative, and all dependence distances can be efficiently computed in polynomial time by an all-pairs shortest-path algorithm [15]. If the SRDF graph is not strongly-connected, there may not be a path between two actors. In such case, we define the dependence distance as infinite.

The linear timing of an SRDF graph can thus be further refined if we take into account the dependence distance. Given a schedule $s(i, k)$ where all firings are self-timed, and actor execution times are given by $t(i, k)$, then, if a new schedule s' is constructed where all firings are self-timed, except for the firing p of actor $n \in V$, which is given by $s'(n, p) = s(n, p) + \delta$ then, for all $i \in V$, $s(i, k) \leq s'(i, k) \leq s(i, k) + \delta$, if $k - p \geq dd(n, i)$ and $s'(i, k) = s(i, k)$, if $k - p < dd(n, i)$, because a change in the execution time of iteration p of i will only affect the firings of i from iteration $dd(n, i) + p$ on.

4.5 Strict Periodicity on a Self-Timed Schedule

There are situations where it is essential to guarantee that an actor has a strictly periodic behavior. For instance, an audio output should not experience any hiccups due to the aperiodic behavior caused by either the initial transition phase of the STS or by the variation of the execution time for different firings of the same actor.

There are two circumstances that make it difficult for an actor to be able to execute in strict periodicity within a self-timed implementation. One is the variation of execution times of its predecessors. Another is the transient phase of the self-timed execution.

Consider the graph in Fig. 4.2a. Actor B represents a strictly periodic external source. Lets say we want to guarantee that actor D executes in a strictly periodic regime. The worst-case self-timed schedule for this graph (up to $t = 8$) is depicted in Fig. 4.2b. After three firings of D, a required input from actor C is still not readily available, and D must wait for one cycle. After this, D executes periodically. The example employs an actor, C, that allows for self-concurrency (multiple firings of C can occur simultaneously), but a similar behavior would be observed for a graph where C is replaced by a chain of four actors with execution time of one, and each with a self-edge with one delay, i.e. the four actors do not allow self-concurrent firings.

A solution to the problem of making D fire in a strictly periodic regime can be found by delaying for some time the first firing of D relative to the first firing of the source (at time 0), but how much time must it be? If we delay the first firing of D for too much time, the strictly periodic execution of the source is violated. For instance, in Fig. 4.2c the first firing of D (D1) is postponed to occur at time $t = 4$. In this case, the external source B will not be able to execute in strict periodicity, because its fifth firing is dependent on the first firing of D.

In this case, a correct solution would be to delay the first execution of D to $t = 2$. This is depicted in Fig. 4.2d. Delaying the first execution of D to $t = 3$ would also work.

There is also the problem with variable execution times. In Fig. 4.2e we depict the case where D1 and D2 execute in less than the maximum execution time of D. Since the token required from B2 is already present, D2 can execute earlier, resulting in aperiodic behavior.

Lets now look for a general solution to the problem of guaranteeing strictly periodic behavior of an actor within a self-timed execution of an SRDF graph. We have already established that for any given period $T \geq \mu(G)$, it is possible to generate an SPS such that all actors are strictly periodic. On the other hand, we know that, in an STS, start times can only be equal or earlier than in an SPS with the same (or a longer) period, i.e.

$$s_{\mathrm{STS}}(i,k) \leq s_{\mathrm{SPS}}(i,k) = s_{\mathrm{SPS}}(i,0) + T \cdot k. \qquad (4.19)$$

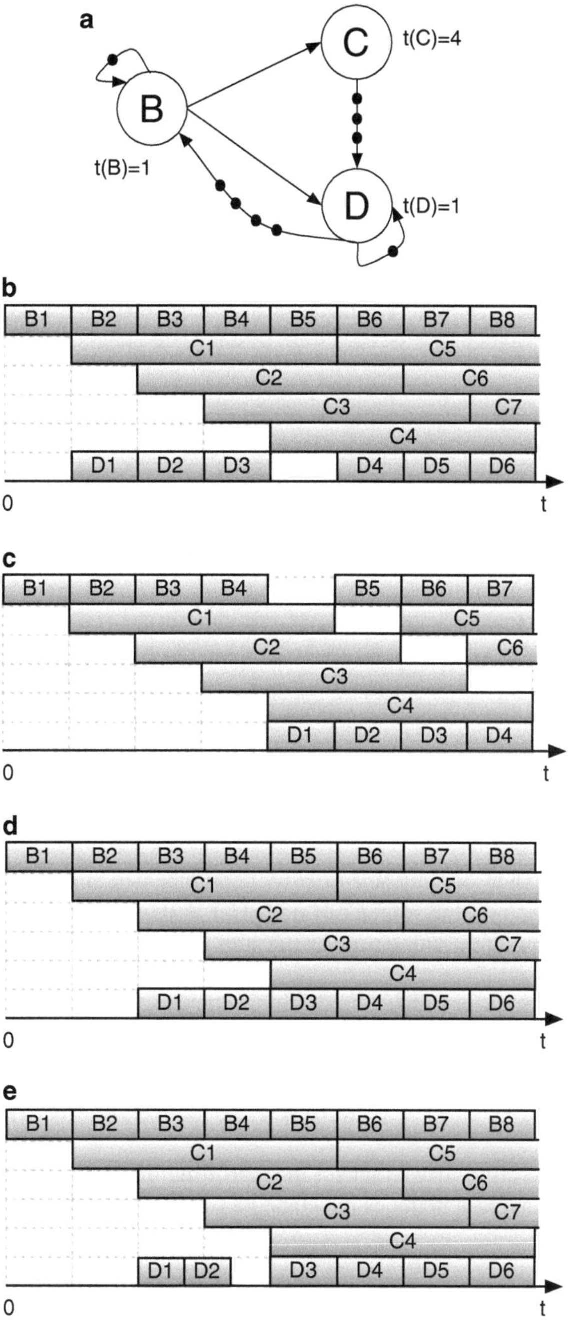

Fig. 4.2 A SRDF graph (**a**), its worst-case STS (**b**), a worst-case STS with D1 forced to start at t = 4 (**c**), a worst-case STS with D1 forced to start at t = 2 (**d**), and a STS with D1 forced to start at t = 2, where D has variable execution time

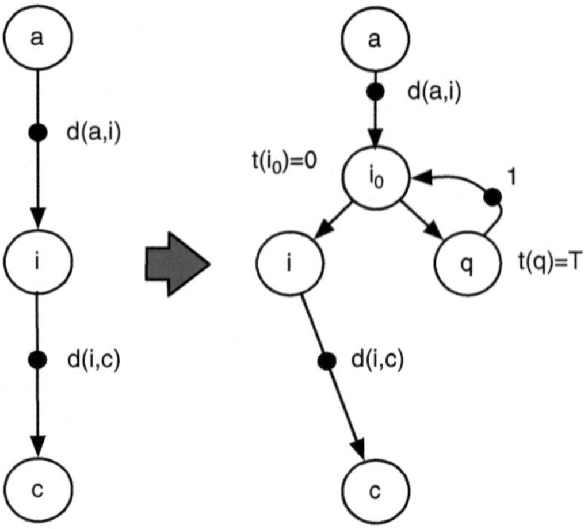

Fig. 4.3 Adding actors and arcs to enforce a minimum time interval T between successive firings of actor i

Assume that we force only a minimum time interval of T between successive starts of an SRDF actor by introducing two additional actors, i_0 and q, with constant execution times $t(i_0) = 0$ and $t(q) = T$. All input edges to i are re-routed to i_0. We further insert edges (i_0, i) with $d(i_0, i) = 0$, (i_0, q), with $d(i_0, q) = 0$ and (q, i_0), with $d(q, i_0) = 1$ as depicted Fig. 4.3, for an actor i with an input edge coming from an actor a, and an output edge going to actor c. Actor i_0 merely forwards the consumed tokens to its output edges (notice that it must forward to i all the inputs from all its input edges). The implementation of actor q can be easily accomplished with a programmable timer. Notice that the inclusion of the extra actors only introduces one cycle with cycle mean T, allowing for our required period.

Then, for any self-timed schedule, i_0 can only be fired once every period T, and, since i_0 is the source of i and $t(i_0) = 0$, all start times of i in a self-timed execution occur at the same time as the start time of i_0 for the same iteration and it holds that:

$$s_{\text{STS}}(i,k) \geq s_{\text{STS}}(i,k-1) + T \Rightarrow s(i,k) \geq s_{\text{STS}}(i,0) + T \cdot k. \tag{4.20}$$

From (4.19) and (4.20) it follows that

$$s_{\text{STS}}(i,0) + T \cdot k \leq s_{\text{STS}}(i,k) \leq s_{\text{SPS}}(i,0) + T \cdot k. \tag{4.21}$$

Therefore, if we can set the start time of the first firing of the self-timed schedule such that it matches the static periodic schedule, i.e.

$$s_{\text{SPS}}(i,0) = s_{\text{STS}}(i,0), \tag{4.22}$$

we can make it such that

$$s_{\text{STS}}(i,k) = s_{\text{STS}}(i,0) + T \cdot k \tag{4.23}$$

What does this imply? That if we fix the start time of its first firing such that the condition in Eq. 4.22 holds for at least one ROSPS of G, we can guarantee i to execute in a strictly periodic fashion, independently of any timing variations that occur in the rest of the graph.

We do not have to enforce an exact initial start time, but guarantee that $s(i, 0)$ is equal to any of the admissible $s_{\text{ROSPS}}(i, 0)$. This means that $s(i, 0)$ must be between its earliest and latest start times in admissible ROSPS schedules – any value in this interval is valid since linear programs have a convex solution space, i.e. any point on the line between two points that meet the linear constraints must also meets the linear constraints. These earliest and latest start times can be computed by finding two ROSPS via LP formulations: one that minimizes i's start time, and another one that maximizes it.

However, if no other condition is imposed, the LP formulation for maximum start time would always result in an unbounded solution, as the start times of all actors can always be shifted to a later time to produce a feasible SPS schedule. In practice, an external source cannot be controlled. In most cases, it is the first firing of the external source that defines the beginning of execution, $t = 0$. This means that in such a case an extra restriction must be placed to the LP: the start time of the first firing of the external source must be equal to 0.

In the implementation, actor i must wait for a time equal to the computed minimum $s(i, 0)$ before firing the first time. After this, the actor may need a local timer (the implementation of actor q) that enables its execution every T units of time, and releases outputs of the previous iteration, such that it exhibits a constant execution time. Essentially, we can statically schedule all firings of one actor (relatively to the start of time $t = 0$) while allowing the rest of the actors to still execute in a self-timed manner and guaranteeing that all data flow dependencies are satisfied.

4.6 Latency Analysis

We will now proceed to present techniques for latency analysis of static DF graphs. We will start by defining latency.

4.6.1 Definition of Latency and Maximum Latency

Latency is the time interval between two events. We measure latency as the difference between the start times of two specific firings of two actors, i.e.:

$$L(i, k, j, p) = s(j, p) - s(i, k). \tag{4.24}$$

Where i and j are actors, p and k firings. We say that i is the source of the latency, and j is the sink.

Typically, since our data flow applications execute for very long sequences of input data, we are interested in iterative latency requirements. That is, we can define that between the pth firing of actor i in any given iteration k and the qth firing of j in iteration $k+n$, where n is a fixed iteration distance, a maximum latency interval is preserved:

$$\begin{aligned}\hat{L}(i,p,j,q,n) &= \max_{k\geq 0} L(i, r(i)\cdot k + p, j, r(j)\cdot(k+n)+q) \\ &= \max_{k\geq 0}(s(j,r(j)\cdot(k+n)+q) - s(i,r(i)\cdot k + p)) \quad (4.25)\end{aligned}$$

with $0 \leq p < r(i)$ and $0 \leq q < r(j)$. Please note that we use $\hat{L}$ to signify the maximum value of the latency across all iterations.

To make the following discussion simpler, we will restrict it to SRDF graphs, where the p and q firing numbers relative to the start of an iteration can be omitted since they are always equal to 0:

$$\hat{L}(i,j,n) = \max_{k\geq 0} L(i,k,j,k+n) = \max_{k\geq 0}(s(j,k+n) - s(i,k)). \quad (4.26)$$

Notice that any latency constraint of the type of Eq. 4.25 can be converted directly into a constraint of the type of Eq. 4.26 in the SRDF-equivalent graph, by applying the relation between MRDF actors and their SRDF copies.

Self-timed scheduling with variable execution times makes latency analysis difficult. The problem is that, while it is easy to find an upper bound for $s(j,k+n)$ using the relations between STS, WCSTS and SPS that we developed in Sect. 4.2, it is still difficult to find a lower bound for $s(i,k)$. In many cases, however, the best-case execution time of the source can be inferred. The simplest case happens if the job has a strictly periodic external source. We will start by analyzing that case.

4.6.2 Maximum Latency from a Periodic Source

The start times of a periodic source are given by:

$$s(i,k) = s(i,0) + T\cdot k. \quad (4.27)$$

Note that the earliest possible value of $s(i,0)$ is given by the WCSTS of the first iteration. Also, note that the period of the execution of the graph is imposed by the source, i.e. because the source executes with a period of T, the period of execution of the graph is lower bounded by the period of the source. If on the other hand, the graph has a longer period, then it cannot keep up with the source, and infinite token accumulation on some buffer will happen for the WCSTS. Therefore, we will perform latency analysis under the assumption that $\mu(G) = T$. Because of Theorem 4.4, the start times of j executing in STS are bounded by the start time of any ROSPS schedule, i.e.

$$s_{\text{STS}}(j,k+n) \leq \check{s}_{\text{ROSPS}}(j,0) + \mu(G)\cdot(k+n), \quad (4.28)$$

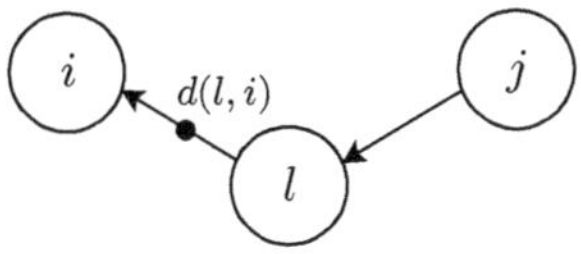

Fig. 4.4 Modeling a latency constraint in an SRDF graph

where $\check{s}_{\text{ROSPS}}(j, 0)$ represents the smallest start time of j in an admissible ROSPS. Equation 4.27 gives us an exact value of $s(i, k)$, while (4.28) gives us an upper bound on $s_{\text{STS}}(j, k + n)$. By taking the upper bound for $s_{\text{STS}}(j, k + n)$, and the lower bound for $s(i, k)$, we get:

$$\hat{L}(i, j, n) = \max_{k \geq 0}(s_{\text{STS}}(j, k + n) - s(i, k)) \leq \check{s}_{\text{ROSPS}}(j, 0) - s(i, 0) + \mu(G) \cdot n. \quad (4.29)$$

Therefore, we can determine the maximum latency from a periodic source just by calculating an ROSPS with the earliest start time j and a WCSTS for the earliest start time of i.

4.6.3 Modeling Latency Constraints from a Periodic Source

We can also represent the latency constraint in terms of a throughput constraint. This is useful when employing an MCM algorithm to check for constraint violation.

We add to the graph an actor l with constant execution time $t(l)$, an edge (j, l) and an edge (l, i) with $d(l, i) \geq 1$ as shown in Fig. 4.4. Actor l does not have a self-edge. The period of the source actor i is $\mu(G)$.

Modeling a latency constraint in this way is only possible between actors with equal repetition vector entries, since we cannot have arcs between specific firings of actors. However, if such model is required, one can always convert the MRDF graph onto its equivalent SRDF.

We will assume, for simplicity sake, that the sink actor of the latency constraint has a constant execution time. At the end of this section we will explain how to handle a sink actor with variable execution time. For this model, it holds that

$$s(i, k + d(l, i)) \geq s(j, k) + t(l) + t(j). \quad (4.30)$$

By replacing (4.5) in (4.30) we obtain:

$$s(j, k) - s(i, k) \leq \mu(G) \cdot d(l, i) - t(l) - t(j) \Leftrightarrow \hat{L}(i, j, 0) \leq \mu(G) \cdot d(l, i) - t(l) - t(j). \quad (4.31)$$

By setting adequately the values of $t(l)$ and $d(l, i)$ we effectively model a latency constraint in terms of the throughput: an infringement of the latency constraint will be detected as an increase of $\mu(G)$, that is, an infringement of the minimum throughput constraint. The parameters can be set for any values of $d(l, i)$ and $t(l) = \mu \cdot d(l, i) - L - t(j)$, as long as $t(l), d(l, i) \geq 0$, but choosing a low value for $d(l,i)$ may constrain the pipelined execution of the iterations of the graph, for analysis purposes. Actor l and edges (j,l) and (l,i) do not need to have any equivalent in the implementation.

If j does not have a constant execution time, then we need one more new actor to model the latency constraint. Lets call this actor j', with execution time $t(j') = 0$. All input arcs of j are redirected to j'. We then add a delay-less edge from j' to j, and build the latency model as described above with j' as the sink, instead of j.

4.6.4 Maximum Latency from a Sporadic Source

In reactive systems, it frequently happens that the source is not strictly periodic, but produces tokens sporadically, with a minimal time interval μ between subsequent firings. Typically, a maximum latency constraint must be guaranteed. This will be the case, for instance, for the WLAN receiver we will show in Chap. 5. It is easy to see that in such case, the MRDF graph has to sustain a throughput of $1/\mu$ in order to guarantee that it cannot be overran by such a source, operating at its fastest rate. This means that the MCM of the graph, $\mu(G)$, is such that $\mu(G) \leq \mu$. Our proof relies on the possibility of bounding the self-timed behavior of a graph by a static periodic schedule with period μ, which is possible as long as $\mu(G) \leq \mu$. In the following text, we will assume for simplicity that $\mu(G) = \mu$, without loss of generality.

In this section, we derive the maximum latency of an actor relatively to a sporadic source. First, we define a sporadic source more formally.

Definition 4.3. A source is sporadic if it fires with a forced schedule s' such that $s'(i,k) \geq s'(i,k-1) + \mu$.

Furthermore, we introduce a strictly periodic reference schedule s of source i with period μ, i.e.

$$s(i,k) = s(i,0) + \mu \cdot k. \tag{4.32}$$

We define $\delta(k)$ as

$$\delta(k) = s'(i,k) - s(i,k), \tag{4.33}$$

that is, delta is the difference between the actual time at which the sporadic source has fired and the time at which our strictly periodic reference schedule of source i fires.

Lemma 4.1. *If a source is sporadic then δ can only increase with every subsequent iteration:*

$$\delta(k+1) \geq \delta(k). \tag{4.34}$$

Proof. We replace the definition of δ in (4.34):

$$(s'(i,k+1) - s(i,k+1)) - (s'(i,k) - s(i,k)) \geq 0. \tag{4.35}$$

As $s(i,k+1) = s(i,k) + \mu$, Eq. 4.35 becomes:

$$s'(i,k+1) - s'(i,k) \geq \mu, \tag{4.36}$$

which is true by hypothesis, since our source is sporadic. □

Theorem 4.6. *Consider an SRDF graph $G = (V, E, t, d)$ and a schedule s' of that graph where all start times are self-timed, except for an external sporadic source n whose forced schedule behaves according to Definition 4.3, with interval μ, and all execution times are worst case. Consider also the schedule s where the sporadic source is replaced by a periodic source $s(i,k) = s(i,0) + \mu$, with $s(i,0) = s'(i,0)$, and where the firings of all other actors are self-timed. Then, for all actors $j \in V$, it holds that:*

$$s'(j,k) \leq s(j,k) + \delta(k - dd(n,j)). \tag{4.37}$$

Proof. For any actor other than the source, all start times in both schedules are self-timed. It therefore holds for any actor $j \in V$ other than the source that

$$s'(j,k) = \max_{(x,j)\in V} (s'(x,k - d(x,j)) + t(x)) \tag{4.38}$$

and

$$s(j,k) = \max_{(x,j)\in V} (s(x,k - d(x,j)) + t(x)). \tag{4.39}$$

If we assume that the theorem applies to all the firings that precede the (j,k) firing, then $s'(x,k) \leq s(x,k) + \delta(k - dd(n,x))$, and it is true that

$$s'(j,k) \leq \max_{(x,j)\in V} (s(x,k - d(x,j)) + t(x) + \delta(k - d(x,j) - dd(n,x))). \tag{4.40}$$

Now, recall that $dd(n,x)$ is the shortest path in delays from n to x. Through the triangle inequality of shortest paths [15], we know that

$$dd(n,x) + d(x,j) \geq dd(n,j), \tag{4.41}$$

for all x. Therefore, recalling Lemma 4.1, it is true that

$$\delta(k - dd(n,x) - d(x,j) \leq \delta(k - dd(n,j)), \tag{4.42}$$

and therefore

$$\max_{(x,j)\in V}(s(x,k-d(x,j))+t(x)+\delta(k-d(x,j)-dd(n,x))) \\ \leq \max_{(x,j)\in V}(s(x,k-d(x,j))+t(x)+\delta(k-dd(n,j))). \tag{4.43}$$

Then, because of the inequality in Eq. 4.40 we can write

$$s'(j,k) \leq \max_{(x,j)\in V}(s(x,k-d(x,j))+t(x)+\delta(k-dd(n,j))). \tag{4.44}$$

and since $\delta(k-dd(n,j))$ is constant inside the max expression we can yank it out to have

$$s'(j,k) \leq \max_{(x,j)\in V}(s(x,k-d(x,j))+t(x))+\delta(k-dd(n,j)). \tag{4.45}$$

Since $s(j,k) = \max_{(x,j)\in V}(s(x,k-d(x,j)+t(x))$

$$s'(j,k) \leq s(j,k)+\delta(k-dd(n,j)) \tag{4.46}$$

□

We can now inspect the latency between an actor j and an external sporadic source i, at an iteration distance of m. The definition of the maximum latency is:

$$\hat{L}(i,j,m) = \max_{k\geq 0}(s'(j,k+m)-s'(i,k)) \tag{4.47}$$

It follows directly from Lemma 4.1 that $\max_{p\leq k}\delta(p) = \delta(k)$. It follows from Theorem 4.6 that:

$$s'(j,k+m) \leq s(j,k+m)+\delta(k+m-dd(i,j)) \tag{4.48}$$

We know that the maximum start time in an STS is not later than the earliest possible start time in an SPS with period μ – which always exists, since $\mu(G) \leq \mu$ – thus

$$s(j,k+m) \leq \check{s}_\mu(j,k+m) \tag{4.49}$$

where $\check{s}_\mu$ is a static periodic schedule of G with period μ.

Given 4.49 and by definition of SPS, we can rewrite 4.48 in:

$$s'(j,k+m) \leq \check{s}_\mu(j,0)+\mu\cdot(k+m)+\delta(k+m-dd(i,j)). \tag{4.50}$$

Equation 4.47 implies that, it holds for all $k \geq 0$ that

$$\hat{L}(i,j,m) \leq s'(j,k+m)-s'(i,k). \tag{4.51}$$

Now, by replacing in this inequality the expressions we obtained for an upper and a lower bound, respectively, for $s'(j, k+m)$ and $s'(i, k)$:

$$\hat{L}(i, j, m) \leq \check{s}_{\mu}(j, 0) + \mu \cdot (k+m) + \delta(k+m-dd(i, j)) - s(i, 0) - \mu \cdot k - \delta(k).$$

We then simplify the right-hand-side expression to obtain:

$$\hat{L}(i, j, m) \leq \check{s}_{\mu}(j, 0) + \mu \cdot m + \delta(k+m-dd(i, j)) - s(i, 0) - \delta(k). \quad (4.52)$$

For $m \leq dd(i, j)$, since $\delta(k) \geq \delta(k+m-dd(i, j))$, we can write:

$$\hat{L}(i, j, m) \leq \check{s}_{\mu}(j, 0) - s(i, 0) + \mu \cdot m.$$

The latency $\hat{L}(i, j, m)$ is not defined for $m > dd(i, j)$ because the start time of execution $k+m$ of j is dependent on the start time of execution $k+1$ of i, which is unbounded, since the maximum difference between $s(i, k+1)$ and $s(i, k)$ is undefined for a sporadic source.

When defined, the latency $\hat{L}(i, j, n)$ with a sporadic source has the same upper bound as the latency for the same source, sink and iteration distance in the same graph with a periodic source with period μ.

4.6.5 Maximum Latency from a Bursty Source

We characterize a bursty source as a source that may fire at most n times within any T time interval, with a minimal Δt interval between consecutive firings. A job that processes such a source must have $\mu(G) \leq T/n$ to be able to guarantee its processing within bounded buffer space. Moreover, if $\mu(G) \leq \Delta t$, then we have the previous case, i.e., maximum latency from a sporadic source. If $\mu(G) \geq \Delta t$ then latency may accumulate over iterations, as the job processes input tokens slower than the rate at which they arrive. The maximum latency must occur when the longest burst occurs, with the minimum interval between firings of the source, that is a burst of n tokens with Δt spacing. Because of monotonicity, making the source execute faster cannot make the sink execute slower, but it also cannot guarantee that it executes faster.

As depicted in Fig. 4.5, the tokens of the bursty source i will arrive earlier than for the periodic source i'. Therefore, iteration $n-1$ after the beginning of the burst (iteration 0) happens the earliest time $s(i, n-1) = s_{\text{ROSPS}}(i, 0) + (n-1) \cdot \Delta t$. The firing $n-1$ of j happens the latest at $s(j, n-1) \leq \check{s}_{\text{ROSPS}}(j, 0) + (n-1) \cdot \mu(G)$. Therefore, a bound on the maximum latency is given by:

$$\hat{L}(i, j, n) \leq \check{s}_{\text{ROSPS}}(j, 0) - s_{\text{ROSPS}}(i, 0) + (n-1)(\mu(G) - \Delta t) \quad (4.53)$$

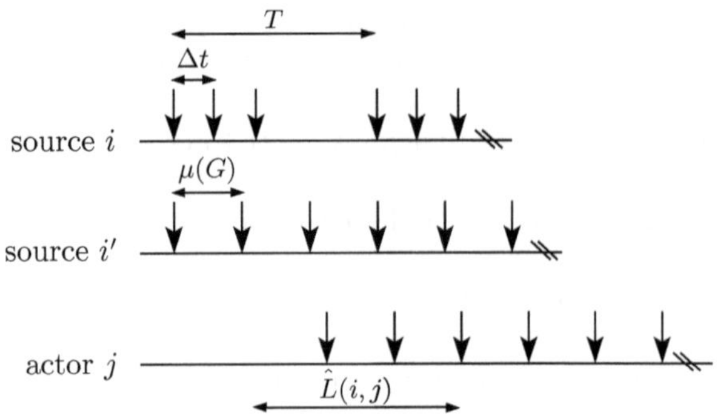

Fig. 4.5 Arrival times of tokens of a bursty source relatively to strictly periodic source

4.7 Further Reading

In [81], latency is defined as the time elapsed between periodic source and sink execution. This book also shows how this can be calculated by simulation of the worst-case self-timed schedule of the job graph. This method requires simulation of the job graph, which is in general intractable, even for Single Rate Data Flow graphs. Its definition of latency is not as general as ours, since it only works if there is at least one delay-less path between source and sink. Moreover, it is only applicable if the external source is strictly periodic. In [24], a method is proposed for measuring the minimal achievable latency of a self-timed implementation of a MRDF graph. This work does not address the problem of bounding the maximum latency.

In [27], latency and buffer sizing are studied in the context of PGM graphs, which are comparable in expressivity to MRDF graphs. The analysis done in this work, however, limits itself to graphs with chain topology. Moreover, [27] does not allow for feedback loops, does not model inter-processor communication, requires EDF scheduling and a strictly periodic source.

The event model used in the SYMTA/S tool [45] cannot cope with critical cycles (i.e., they are not taken into account), and latency can only be measured as a result of a complete mapping, never taken into account as a constraint during the mapping processes, as our analysis method allows.

In [21], a proof is given that a periodic linear bound exists on consecutive production times of tokens that occupy the positions of initial tokens for a MRDF graph. These are equivalent to the linear bounds that can be obtained with our method for the firing times of the producers of such tokens. This paper also proposes a way of computing these bounds through a simulation-based technique.

4.8 Conclusion

In this chapter, we have presented temporal analysis methods for static data flow, that are derived from the monotonic property of MRDF graphs and especially on the relation between self-timed and periodic schedules. We use this relation to reason not only about average throughput but also about maximum latency and periodic behavior. These techniques allow us to provide conservative linear bounds for all firings of actors of the MRDF graph – even during the transient phase of self-timed execution – and bound the effect of an increase of the worst-case execution time of an actor on the behavior of the graph.

Chapter 5
Resource Modeling and Compile Time Scheduling

In our approach to Software-Defined Radio, a radio transceiver is a concurrent software program expressed by means of a data flow graph. In this chapter, we show how such a data flow graph can be partially scheduled at compile time to a heterogeneous multiprocessor platform, in such a way that guarantees of Hard Real-Time performance can be given. The scheduling decisions taken at compile time are not complete, since, as already explained in Chap. 2, the assignment of actors to processors in the physical platform must be done at runtime, to preserve the flexibility in the mapping that allows for multiple instances of multiple transceivers to run simultaneously. Nonetheless, we take as many scheduling decisions as possible at compile time to decrease the complexity of the runtime mapping algorithm that assigns hardware resources to each transceiver instance.

Accordingly, we take at compile time all the scheduling decisions that require complex end-to-end temporal analysis of the graph. These include the computation of buffer sizes, clustering of actors and static ordering of actors per cluster, and the resource requirements towards the local runtime schedulers of each processor.

Our strategy is to take the original actors of the data-flow graph and group them in clusters of statically-ordered actors. Each cluster will be assigned as one entity to one processor by the online processor allocation algorithm. We then determine for each of these clusters the amount of processor resources, including scheduler settings for a Time Division Multiplexing (TDM) or a Non-Preemptive Non-Blocking Round Robin (NPNBRR) scheduler, and for all inter-cluster channels all buffer capacities, in such a way that they are sufficient for the transceiver to meet its end-to-end Hard Real-Time (HRT) requirements.

We form clusters by mapping actors to virtual processors, and statically order all the actors mapped to each given virtual processor. Each virtual processor represents a fragment of the resources of a physical processor in the target platform. The designer must specify to the scheduler a target platform file containing a list of all virtual processors and their resources. We assume that all virtual processors of a given type can be mapped to a physical processor of the same type. For the sake

O. Moreira and H. Corporaal, *Scheduling Real-Time Streaming Applications onto an Embedded Multiprocessor*, Embedded Systems 24, DOI 10.1007/978-3-319-01246-9_5,

of simplicity, we will assume that all physical processors of a given type have the same clock frequency, and the same dynamic scheduler with the same settings.

We have shown in Chaps. 3 and 4 that the self-timed execution of Single Rate Data Flow (SRDF) and Multi Rate Data Flow (MRDF) can be analyzed for worst-case temporal behavior, assuming sufficient buffer sizes. Our compile time scheduling algorithm requires the generation of a data flow graph that models the temporal behavior of the input graph including the effects of each scheduling decision we take. Since scheduling decisions can only decrease the throughput or increase the latency of the graph execution, we can take decisions in incremental steps, and check, at each step, if each new decision still allows for a valid schedule. We can then exhaustively search the tree of scheduling decisions for solutions, backtracking each time an infeasible branch is found.

Our data flow scheduler is designed to schedule single-rate (SRDF), multi-rate (MRDF) and cyclo-static graphs. MRDF and CSDF graphs are first converted onto an SRDF equivalent, but special care must be taken to account for the fact that multiple arcs in the converted graph correspond to a single buffer FIFO. For any actor in the input data flow graph that keeps state over consecutive iterations (or equivalently has a self edge with unitary delay), we will require that the SRDF copies are mapped onto the same processor, to avoid state migration.

We will first explain how the scheduler works for input graphs that are already single-rate (SRDF). We will then dedicate a section to explaining what must be changed in the scheduler to accommodate for graphs that were originally multi-rate (MRDF) or cyclo-static. In Sect. 5.1 we will describe and provide notation for the inputs of our scheduler. In Sect. 5.2 we describe and provide notation for the outputs of our scheduler. In Sect. 5.3, we will show how the compile time scheduling decisions can be modeled conservatively by means of a data flow graph, and how a temporal model for a partial schedule of a graph can be constructed. After this, in Sect. 5.4 we motivate the need for applying a combination of clustering and static ordering with a runtime scheduler. In Sect. 5.5 we define the compile time scheduling problem and propose strategies to tackle it. In Sect. 5.6 we apply our techniques to obtain schedules for a TD-SCDMA receiver and a WLAN receiver, that must share resources of a multi-processor platform. In Sect. 5.7 we discuss related work and in Sect. 5.8 we state our conclusions for this chapter.

5.1 Scheduler Inputs

Our scheduler requires three inputs: the task graph of the transceiver, a description of the target platform onto which the transceiver is going to be mapped, and a set of timing requirements. Furthermore, the user can optionally specify additional constraints to the mapping. In the following sections we detail these inputs.

5.1.1 Target Platform

We model the hardware platform to which the application graph is going to be mapped as a set of virtual processors. Each virtual processor represents a quantity of processing resources made available by the physical hardware platform. We chose to allocate resources of virtual processors at compile time because this still allows the binding of virtual to physical processors to be done at runtime, resulting in more mapping flexibility. Each virtual processor may represent all or a fraction of the resources of a physical processor. For example, we may map our application to a family of hardware platforms that include two to four ARM processors. The virtual platform described as input to the compile time scheduler can contain two (virtual) ARM processors. At runtime, the resource manager may decide to allocate each of the virtual processors to any two of the four physical processors, if the particular instance of the hardware platform contains four ARM processors.

A virtual target platform is described by a set of virtual processors Π. The processors are described by a tuple $(\tau, sched, P, \hat{S})$. The valuation $\tau : \Pi \rightarrow T$, where T is the set of processor types, represents the type of the processor (for instances, an ARM or an EVP). The valuation $sched : \Pi \rightarrow$ *{tdm, rr, off}* represents the type of the dynamic scheduler that is running on the processor. In the current implementation, this type can be *tdm*, if the processor implements a TDM scheduler, *rr* if the processor implements a NPNBRR scheduler, or *off*, if the processor has no dynamic task scheduler running. The valuation $P : \Pi \rightarrow \mathbb{N}_0$ represents the period of the scheduler (in a round-robin scheduler, this is the sum of the worst-case execution times of actors that can at any time be mapped onto the processor). The valuation $\hat{S}$ represents the maximum amount of processor resources that a single job can use per period. Therefore it must always hold that $\hat{S} \leq P$. Furthermore, we assume that all processors of a given type have the same value for *sched* and P, and we can therefore refer to the valuations $sched : T \rightarrow$ *{tdm, rr, off}* that, given a processor type, returns the type of scheduler used in those processors, and valuation $P : T \rightarrow \mathbb{N}_0$ that, given a processor type, returns the period of the scheduler for that type of processor.

One more observation about the relation between virtual processors and physical processors is needed: two virtual processors can map onto a single physical processor during runtime mapping. If, for example, we have two virtual processors with the same scheduler and period, and both with a maximum utilization lower than half the period, the runtime resource allocator will always be able to map both virtual processors to the same physical processor, if resources are available at the time of the allocation. By deciding how the target platform is described in terms of virtual processors to the compile time scheduler, the designer can effectively control the cluster granularity.

5.1.2 Task Graph

A timed SRDF graph, annotated with processor type per actor and token size per arc, $G = (V, E, d, t, \tau, \beta)$ represents the data flow program that corresponds to a transceiver. V is the set of actors, E is the set of arcs. Each actor in V corresponds to an invocation of a computational task in the input LIME program. We call this graph the **Task Graph** of the application. The $t(i)$ valuation of actors corresponds to the WCETs of the computational task invocation on a specific processor type. We assume that each actor in the task graph can only be mapped to a specific type of processor. The type of processor on which a particular actor runs is given by an extra valuation $\tau : V \to T$. The valuation $\beta : E \to \mathbb{N}_0$ represents the size $\beta(i, j)$ in bytes of each token produced/consumed in arc $(i, j) \in E$.

5.1.3 Timing Requirements

Input timing requirements are both throughput and latency constraints. An actor $i \in V$ may be required to have a minimal guaranteed throughput of $\theta(i)$, and maximum latency requirements can be expressed as described in Chap. 4. The throughput requirements are used to determine a required maximum cycle mean μ_D for the graph. The maximum period has to be such that all nodes meet their throughput requirement and therefore $\mu_D = \min_{i \in V} \theta(i)^{-1}$. Latency constraints are expressed in the temporal analysis graph in terms of μ_D by using the latency modeling technique described in Sect. 4.6.3, on top of the temporal analysis model that will be described in Sect. 5.3.

5.1.4 Additional Constraints

The scheduler accepts a number of additional mapping constraints. These will not be discussed in the rest of this chapter, as adding verification of these constraints to the scheduler is conceptually straightforward, while including them in the remainder of the text will make all explanations and notations much longer. These constraints are the following:

1. **Memory space**: besides cycles, actors consume memory space. Dependent on the processor architecture, these may include several different memory pools, such as Instruction and Data Memory for Harvard architectures, or Instruction, X-Data and Y-Data for Dual-Harvard architectures. The data memories can be further subdivided in read-write, stack and read-only memories, for instances. Our scheduler allows for actors to be annotated with a vector of memory requirements, and virtual processors in the target architecture to be annotated

with a vector of memory provisions. Any valid schedule must respect that the sum of memory requirements of all actors mapped to a given processor is lower than the memory provided for each pool in that processor.
2. **Forced Mapping**: the designer can constrain the mapping of actors to a specific processor in the virtual target platform.
3. **Forced Co-mapping**: the designer can instruct the scheduler to force a set of actors in the task graph to be mapped to the same processor, i.e. the designer imposes that the set of actors belongs to the same cluster.
4. **Forced Ordering**: the designer may specify a order of execution between firings of two actors by adding an extra arc between them.
5. **Buffer size restriction**: the designer can instruct the scheduler to assume that the maximum buffer space between two actors is limited by a given value. This will be further discussed in Sect. 5.3.2.

5.2 Scheduler Output

The output of the compile time scheduler is given as a static order, a slice size per processor, and a buffer capacity per edge in the input graph. The static order is represented by the valuation $so : \Pi \rightarrow \alpha^n$, where α^n is the set of actor sequences of the type, $[i_1, i_2, \ldots i_n]$, where $i_1, i_2, \ldots, i_n \in V$, and the slice size for the cluster is given by valuation $S : \Pi \rightarrow \mathbb{N}_0$. The static order also implies a processor allocation $\pi : V \rightarrow \Pi$, such that, if, for a given $i \in V$, if $i \in so(p) \Longrightarrow \pi(i) = p$, meaning that actor i is mapped to processor p. In a complete compile time schedule, for any actor $i \in V$ there must exist a single processor p such that $i \in so(p)$. The buffer capacity is represented by valuation $b : E \rightarrow \mathbb{N}_0$, and represents the amount of tokens allocated to the edge.

5.3 Modeling Resource Allocation

In order to check whether a particular compile time schedule meets the timing requirements, we generate a temporal analysis graph G' that represents the effects on the temporal behavior of graph G of the compile time schedule described by (so, S). We must be able to conservatively model the effects of actor ordering, local dynamic schedulers, communication channels and buffer sizing. We will discuss first each one of these and then provide a transformation that, given the input data flow graph G, a target platform Π and a (partial) compile time schedule (so, S), generates the temporal analysis graph $G'(G, \Pi, so)$.

5.3.1 Communication Channels

Depending on the target architecture and the level of detail required, communication channels between actors executing in different processors might be modeled in different ways. In [71] and [78] models are derived for the Æthe real network. Many different models for the same network are possible, depending on the level of abstraction. In the remainder of this chapter, we will assume that communication between actors is done via remote stores whose worst case time is accounted for in the execution times of the producing actors. In Chap. 7 we will propose a way of modeling the latency and rate effects of Network-on-Chip communication. The reader is encouraged to also consult [71] and [78] to find more precise and detailed models of communication. Any of these models is parametric, depending on token size and the characteristics of the inter-processor communication hardware.

5.3.2 Buffer Size Restriction

A buffer capacity constraint can be modeled in an MRDF graph by adding to the input graph a back-edge from the consumer of a FIFO to its producer with production/consumption rates that match the consumption/production rates of the forward edge. As the number of tokens in the cycle between producer and consumer can never exceed the number of initial tokens in that cycle, the edge that models the actual data FIFO can never have more then the number of tokens initially placed in the "credits" back edge plus the number of initial tokens in the forward edge. This also means that an actor cannot fire without enough space being available in each of its output FIFOs, which represents the worst-case effect of back pressure.

5.3.3 Task Scheduling

There are two types of task scheduling mechanisms that we are interested in modeling: Compile Time and Runtime Scheduling.

Compile Time Scheduling (CTS) encompasses scheduling decisions that are fixed at compile time, such as static order scheduling.

Runtime Scheduling (RTS) refers to scheduling decisions that cannot be fully resolved at compile time, because they depend on the runtime task-to-processor assignment, which in turn depends on the dynamic job-mix. This is handled by the local scheduling mechanism of the processor. Modeling the worst-case effect of the local scheduler on the execution of an actor is needed to include in the compile time analysis the effects of sharing processing resources among jobs. If the WCET of the task, the settings of the local dispatcher, and the amount of computing resources to be given to the task are known, then the actor execution time can be set to reflect the **worst-case response time** of that task running in that local dispatcher, with that particular amount of allocated resources. In [7], we show how this can be computed for a TDM scheduler and, in [74], for a NPNBRR scheduler. We will present these models in the following subsection and extend them to handle its combination with static order scheduling.

5.3.4 TDM Scheduling

In [7], it has been shown that the effect of TDM scheduling can be conservatively modeled by replacing the worst-case execution time of the actor by its worst-case response time under TDM scheduling. The response time of an actor i is the total time it takes to fire i, when resource arbitration effects (scheduling, preemption, etc.) are taken into account. This is counted from the moment the actor meets its enabling conditions to the moment the firing is completed. Assuming that a TDM wheel period P is implemented on the processor and that a time slice with duration S is allocated for the firing of i, such that $S \leq P$, a time interval equal or longer than $t(i)$ passes from the moment an actor is enabled by the availability of enough input tokens to the completion of its firing. This is due to what can be seen as two different effects of TDM arbitration. The first of this is the **arbitration time**, i.e. the time it takes until the TDM scheduler grants execution resources to the actor, once the firing conditions of the actor are met. In the worst-case, i gets enabled when its time slice has just ended, which means that the arbitration time is the time it takes for the slice of i to start again. If we denote the worst-case arbitration time as r_A then:

$$r_A(i) = P - S. \tag{5.1}$$

The second effect has to do with the fact that the time slice may be too small for the firing of i to be executed in a single slice. We shall call the time between the moment that the actor starts execution and finishes execution the **processing time** of the actor, and denote it by r_P. The time i will take to fire, in the worst-case, is equal to

$$r_P(i) = \lfloor \frac{t(i)}{S} \rfloor \cdot P + (t(i) \bmod S). \tag{5.2}$$

Since $t(i) \text{mod} S = t(i) - S.\lfloor \frac{t(i)}{S} \rfloor$, the previous expression can be re-written as:

$$r_P(i) = \lfloor \frac{t(i)}{S} \rfloor \cdot (P - S) + t(i). \tag{5.3}$$

The total worst-case response time of i is then given by the sum of these two values:

$$r(i) = r_A(i) + r_P(i) = (P - S) \cdot (\lfloor \frac{t(i)}{S} \rfloor + 1) + t(i). \tag{5.4}$$

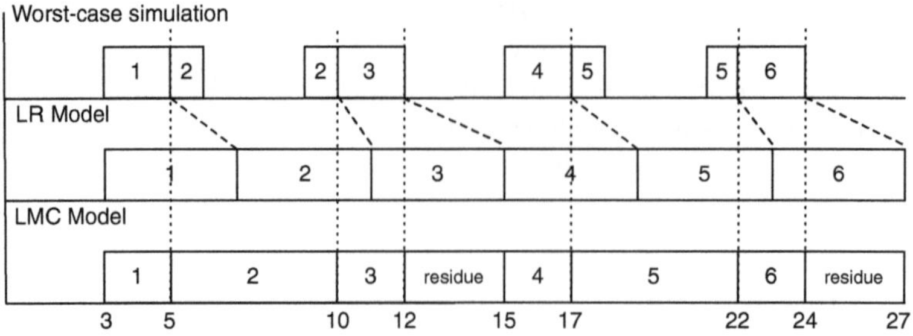

Fig. 5.1 Response times for LR and LMC models compared with actual TDM behavior for a burst

By replacing the $t(i)$s of all actors scheduled under TDM by their worst-case response times, $r(i)$s, temporal analysis of the timed SRDF graph thus obtained will yield the minimum guaranteed throughput that is attainable by such an implementation.

In later work [90], it was shown that a more accurate model of a TDM scheduler can be built by modeling a generic latency-rate server [83] in data flow. The effect on the timing of an actor i with execution time $t(i)$ of executing it under a latency-rate server can be modeled in data flow by replacing the actor by two actors: an actor i_L and an actor i_R. All the input edges of actor i are connected as inputs to i_L, and all output edges of actor i are connected as outputs to the rate actor i_R. Furthermore, an edge (i_L, i_R) with no delays and an edge (i_R, i_R) with one delay are added to the graph. The execution times of the actors in the model are given by

$$t(i_L) = P - S \tag{5.5}$$

and

$$t(i_R) = P.\frac{t(i)}{S}. \tag{5.6}$$

This model is known as the Latency-Rate (LR) model.

5.3.5 *The LMC Model*

The Latency-Rate model can still grossly over-estimate the response time of an actor running on a TDM scheduler. To understand why that is, we will provide an example. Figure 5.1 depicts on the top row the execution of the first six firings of an actor x with an execution time $t(x) = 2$, running on a processor with a TDM schedule with period $P = 6$ and a slice $S = 3$, assuming that input tokens are always available. This first schedule shows the actual simulation of the time wheel, assuming constant execution time and that the first time slice allocated to actor x occurs exactly at the worst case waiting time $P - S = 3$. The schedule in the second row represents the self-timed execution of the LR model for this example, as described in the previous section. As it can be seen, the fact that the LR model uses one single actor to represent the burst behavior of the actor forces it to model the effects of the TDM scheduler such that start times and finish times of consecutive iterations of the actor occur at evenly spaced intervals. Since the model must be conservative, this results in over-estimation of finish times.

A more accurate model can be devised by realizing that, when executing a continuous stream of activations, i.e. a case where the actor is always ready to fire at the end of a previous firing, the service provision of the TDM scheduler to an actor with a fixed execution time follows, as shown in Fig. 5.1, a cyclo-static pattern of behavior: after a number of firings q such that $q \cdot t(x) = p \cdot S$, where $t(x)$ is the worst-case execution time of an actor x mapped to a slice of size S of a TDM scheduler, and p and r are integers, the following q iterations occur such that the distance between start and finish times for any iteration k is the same as for $k\,\%q$. We can thus construct a model that treats each of the q iterations in this cycle separately, thus accurately specifying the worst-case behavior per iteration. This model, which we will call LMC after the names of the authors, has been proven to be conservative in [58].

To describe how the model is constructed, we must introduce a function, also established in [58], that defines the worst-case relation between arrival and finish times of actor firings under TDM scheduling, $F_{P,S,\phi,t} : \mathbb{R} \rightarrow \mathbb{R}$ that returns, for a given TDM period P, slice S, phase ϕ, and actor execution time t, the finishing time $F_{P,S,\phi,t}(\tau)$ for a single iteration arriving at time τ. This time depends on the exact position of the TDM wheel at the time of the arrival of the actor iteration, and on the splitting of execution time t over consecutive TDM slices. The modulo function can calculate the relative position of times in the TDM wheel. For example, $(\phi - \tau)\%P$ denotes the time to the next start of the allotted slice. In the best case, the iteration arrives exactly at the start of the slice (i.e. when $(\tau - \phi)\%P = 0$), and at least $\lceil \frac{t}{S} - 1 \rceil$ full periods are needed to execute the actor, plus a remainder of $t - \lceil \frac{t}{S} - 1 \rceil$. If the iteration arrives outside the slice, the time to the next slice must be added, and if the iteration arrives within the slice but too late to finish the 'remainder', a full $P - S$ is added. Thus, three cases are distinguished:

$$F_{P,S,\phi,t}(\tau) = \begin{cases} \tau + t + \lceil \frac{t}{S} - 1 \rceil (P - S) + (\phi - \tau)\%P \\ \quad ; \text{for } (\phi - \tau)\%P \leq (P - S) \\ \tau + t + \lceil \frac{t}{S} \rceil (P - S) \\ \quad ; \text{for } (\phi + S - \tau)\%P < t - \lceil \frac{t}{S} - 1 \rceil S \\ \tau + t + \lceil \frac{t}{S} - 1 \rceil (P - S) \\ \quad ; \text{elsewhere} \end{cases}$$

Notice that function F gives the finish time of one iteration if the arrival occurs when the TDM scheduler is not busy executing a previous iteration. Notice also that if a burst of i iterations occurs, i.e., in i consecutive iterations each iteration arrives before or at the same time that the previous iteration is completed, the finish time of the burst occurs as if a single firing with a duration of $i \cdot t$ times units has occurred, that is $F_{P,S,\phi,i \cdot t}(\tau)$ where τ is the start time of the first iteration in the burst.

As mentioned above, the crucial observation for this approach is that, if S and τ are rational numbers, it is possible to pick $q, r \in \mathbb{N}$ such that $q \cdot t(x) = r \cdot S$. This is a hyper-period in which q consecutive iterations will need exactly r periods to finish execution, provided one starts within the allotted slice.

The LMC dataflow model details the timing of $q + 1$ consecutive iterations. Each iteration becomes a node in the SRDF graph component $G_j = (V, E, d, t)$ which will replace the node of the TDM-scheduled task x in the original application graph. Adding the initial latency, the graph will have $q + 2$ nodes in total, so the set V of actors of the model is given by:

$$V = \{x_i \mid 0 \le i \le q + 1\},$$

where actor x_0 denotes the initial 'latency' or 'black-out' of TDM, and the other actors form a cycle:

$$\begin{aligned} E = &\{(x_{i+1}, x_i) \mid 1 \le i \le q\} \cup \{(1, q + 1)\} \\ &\cup \{(x_0, x_i) \mid 1 \le i \le q + 1\} \end{aligned}$$

Each edge of the cycle obtains an initial token, except the edges from x_0 and the edge from x_1 to x_{q+1}:

$$d(x_i, x_j) = \begin{cases} 0 & ; \text{for } i = 0 \\ 0 & ; \text{for } i = 1 \wedge j = q + 1 \\ 1 & ; \text{elsewhere} \end{cases}$$

Each actor in the cycle has an execution time corresponding to the distance between start and finish time of an actor iteration arriving at that point in the cycle, taking into account that actor x_0, like the latency actor in the LR model, has an execution time of $P - S$ to compensate for worst-case phasing (i.e. we assume the burst starts at the worst possible position of the time wheel).

$$t(x_i) = \begin{cases} P - S & ; \text{for } i \in \{0, q + 1\} \\ F_{P,S,\phi,t(x_i)}(\phi) - \phi & ; \text{for } i = 1 \\ F_{P,S,\phi,i \cdot t(x_i)}(\phi) - F_{P,S,\phi,(i-1) \cdot t(x_i)}(\phi) & ; \text{for } 1 < i < q + 1 \end{cases}$$

As an example, we show the model corresponding to the case mentioned above where $t(x) = 2$, $P = 6$, $S = 3$, in Fig. 5.2. The minimal value for parameter q is given by $S/gcd(S, t(x)) = 3$ resulting in a model with $q + 2 = 5$ actors, with execution times given by the $t(x_i)$ expression presented above.

The LMC model has been proven in [58] to be conservative towards the actual execution of a TDM scheduler. The same paper also proves that the response

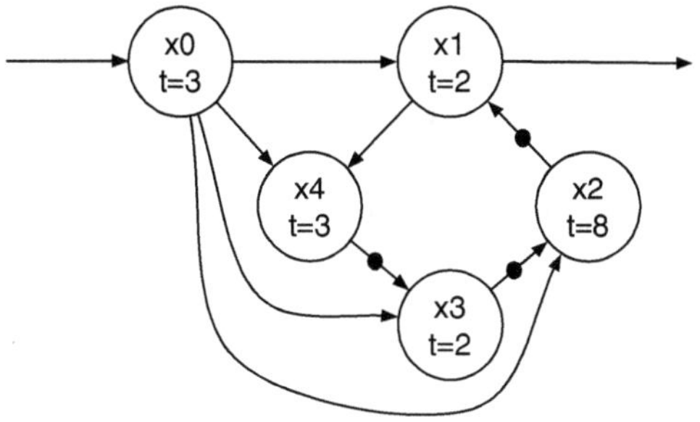

Fig. 5.2 LMC response model for $t(x) = 2$, $P = 6$, $S = 3$

behavior is always equal or better than obtained by the LR model. Furthermore, it has been shown to lead to resource allocation improvements (a 20 % improvement was reported in [58] for a simple application example), when compared with the LR model. This model, however, has some disadvantages. The first is that it may require many actors and edges to represent the temporal response of a single actor mapped to a TDM wheel. There are $q + 2$ actors and $2q + 2$ edges where the smallest q, for integer S and $t(x)$, is given by $S/gcd(S, t(n))$. This can be a large number, resulting in large graphs to be analysed by MCM algorithms whose complexity depends on the number of actors and edges. A second disadvantage is that the model assumes the worst-case execution time $t(n)$ is the same for successive activations. This makes it difficult to apply the LMC model when several actors with different worst-case execution times are statically ordered in one TDM slot – we will discuss why we are interested in such arrangement in Sect. 5.4. Such situation can still be approximated by assuming the worst-case of all worst-case execution times for actors sharing the TDM slot, but such simplification will lead to overestimation. This can offset the advantages over the LR-model, which can be easily adapted to handle combined static-order and TDM scheduling (as we shall see in Sect. 5.3.8). Due to this restriction – which may be lifted soon as there is ongoing research on this topic – and for simplicity's sake, we will, for the remainder of this book, continue to use exclusively the LR model. The reader should be aware that at least in the case where a single SRDF actor is mapped to a TDM resource, it could be advantageous to apply the LMC model instead.

5.3.6 NPNBRR Scheduling

In a Non-Preemptive Non-Blocking Round-Robin (NPNBRR) Scheduler, all clusters assigned to the same processor are put in a circular scheduling list. The runtime scheduler goes through this list continuously. It picks an actor from the list and tries to execute it. The actor (or the scheduler, depending on the implementation) checks for input data and output space availability. If there are sufficient input data tokens available in all input FIFOs and output space available in all output FIFOs such that the actor can consume and produce tokens according to its firing rules, the actor executes until the firing is over, if not, the actor is skipped. The process is repeated for the next actor in the circular scheduling list, and so on.

The worst-case arbitration time of an actor is given by the sum of the execution times of all other actors mapped to the same NPNBRR-scheduled processor. The processing time is equal to the actor's execution time, since there is no preemption. The total response time is therefore equal to the sum of the execution times of all actors mapped to the NPNBRR-scheduled processor.

As for the representation of scheduler settings for an NPNBRR scheduler, we will take the maximum allowed sum of worst-case execution times of tasks per turn of the Round-Robin wheel as the chosen maximum period P and the maximum sum of worst-case execution times for a given cluster of statically-ordered actors as the maximum slice $\hat{S}(p)$ of the resources of processor p by a task graph G.

5.3.7 *Static Order Scheduling*

A static order schedule of a set of actors $A = \{a_0, a_1, \ldots, a_n\}$ mapped to the same processor is a sequence of execution $so = [a_k, a_l \ldots a_m]$ that generates extra precedence constraints between the actors in A such that from the start of the execution of the graph, a_k must be the first to execute, followed by a_l and so on, up to a_m. After a_m executes, the execution restarts from a_k for the next iteration of the graph.

Any static order imposed to a group of SRDF actors executing on the same processor can be represented by adding edges with no tokens between them. From the last to the first actor in the static order an edge is also added, with a single initial token. This construct reflects the fact that, the graph execution being iterative, when the static order finishes execution for a given iteration, it re-starts it from the first actor in the static order for the next iteration.

Notice that the new edges represent a series of sequence constraints enforced by the static order schedule and do not represent any real exchange of data between the actors. In the data flow diagrams that follow, for ease of reading, every time there is more than one edge from the same source to sink, only one edge with the lowest $d(i, j)$ is represented, as it imposes the tightest sequence constraint between the actors.

5.3.8 *Combining Static Order and TDM*

Let's assume that instead of attributing a separate TDM slice to each actor we attribute a single TDM slice to a group of actors and we statically order these actors. The reasons why we may want to do this are exposed in Sect. 5.4. For now we will focus only on how such a mixed schedule can be modeled for the purpose of worst-case analysis.

Consider a set of actors $\{a_1, a_2, a_3\}$, belonging to the same job, and mapped to the same processor p. Consider that a static order of execution $so = [a_1, a_2, a_3]$ has

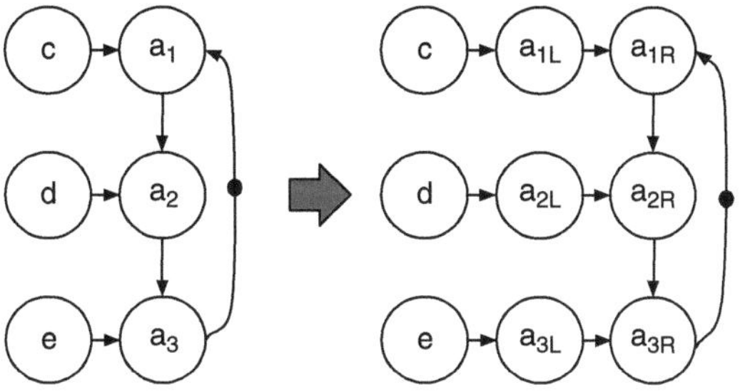

Fig. 5.3 Modeling the combined effect of static order and TDM scheduling

been imposed on them, and that this static order execution is allocated to a time slice of S time units within a time wheel with a period P.

Since the actors are guaranteed to be mutually exclusive because of the static order, it turns out that whenever one of them is activated none of the remaining others is. Thus, we can conservatively assume that the TDM scheduling will affect the timing of executing each actor in the same way as it would if the time slice were allocated exclusively to it. We can therefore model the combined effect of the static order and TDM scheduler by adding to the scheduling analysis graph the edges that represent the imposed static order and replacing each actor by the latency-rate model of its execution under a budget scheduler, as described in Sect. 5.3.4. This is depicted in Fig. 5.3 for three actors a_1, a_2 and a_3, statically ordered and assigned a slice with time S.

5.3.9 *Mixing Static Order and NPNBRR Scheduling*

Modeling the effect on timing of NPNBRR scheduling over a cluster of statically-ordered actors is somewhat similar to the previous case. The main difference is that each actor will run to completion once its input is available, without being preempted by the scheduler.

Consider $C = \{a_1, a_2 \dots a_n\}$, a set of actors belonging to the same job G, and mapped to the same processor p. Consider that a static order of execution $so = [a_1, a_2, \dots a_n]$ has been imposed on them, and that the total sum of execution times of actors in the Round-Robin list of processor p is P. This includes the actors belonging to this cluster and any other clusters that the runtime resource manager has allocated to the same processor. Under these conditions, the first actor in the cluster, a_1, will be able to fire for the first time, in the worst case, after the time at which its first input token has been produced plus an arbitration time equal to $P - \sum_{a_n \in C} t(a_n)$. This time may be used by actors from other jobs running on the same physical processor. Note that actors belonging to the same cluster as a_1 are not accounted for here, since those dependencies are modeled with edges in the graph, and enforce mutual exclusivity. After a_1 finishes firing, a_2 can immediately fire if input data is available. If not, the firing will be delayed,

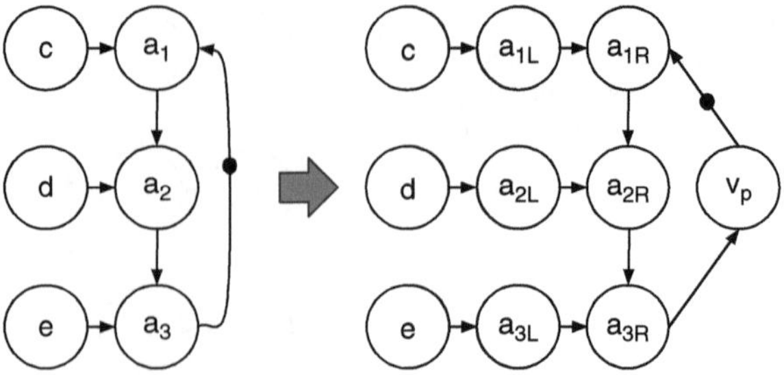

Fig. 5.4 Modeling the combination of static order and NPNBRR

and the Round Robin scheduler moves to the next cluster. Once an input token has become available, the NPNBRR scheduler can take a time at most equal to $P - \sum_{a_n \in C} t(a_n)$ to allow a_2 to fire the actor The same is true for a_3 and all other actors in the cluster. This effect can be modeled by introducing in the analysis model, for each actor $a_i \in C$, a latency actor a_{iL} with $t(a_{iL}) = P - \sum_{a_n \in C} t(a_n)$, adding edges (a_{iL}, a_i), with $d(a_{iL}, a_i) = 0$, and further replacing all edges in the set $(j, a_i) \in E : a_i \in C, j \notin C$, by edges (j, a_{iL}) each with the same delay as the corresponding (j, a_i).

Since each actor in the static order is still limited to execute at most once per period, it occurs that in the worst case, after the last actor has executed, the first can only execute again after a time interval of $P - \sum_{a_n \in C} t(a_n)$ has elapsed, even if input tokens are readily available. This can be modeled by adding an actor with execution time $P - \sum_{a_n \in C} t(a_n)$ at the end of the cluster. Figure 5.4 depicts the conversion to the analysis model for a cluster of three statically ordered actors a_1, a_2 and a_3.

5.3.10 Temporal Analysis Model

Having defined how to model the effects of our compile time scheduling decisions on an SRDF graph, we are ready to present a graph transformation that allows us to obtain a temporal analysis model SRDF graph G', from the knowledge of the original task graph G, the target platform Π, and the compile time schedule (so, S).

We first define the set of edges in G internal to virtual processors:

$$E_{int} = \{(i, j) \in E \mid \pi(i) = \pi(j)\}, \tag{5.7}$$

and the set of edges between virtual processors

$$E_{ext} = \{(i, j) \in E \mid \pi(i) \neq \pi(j)\} = E - E_{int}. \tag{5.8}$$

The reason why we separate the treatment of external and internal edges is that we do not need to assume that internal edges must wait for the latency term in the response model, as both producer and consumer actors must belong to the same cluster, and their dependencies are already modeled in the graph. The analysis model $G'(G, \Pi, so, S, P) = (V', E', t', d')$ is then given by replacing all actors in G for their latency-rate models as described above, adding edges to represent the static ordering, and computing the execution times of these new nodes:

$$
\begin{aligned}
V' &= \{i_L \mid i \in V\} \cup \{i_R \mid i \in V\} \cup \{v_p \mid p \in \Pi\};\\
E' &= \{(i_L, i_R) \mid i \in V\} \cup \{(i_R, i_R) \mid i \in V\}\\
&\quad \cup \{(i_R, j_R) \mid (i, j) \in E_{int}\} \cup \{(i_R, j_L) \mid (i, j) \in E_{ext}\}\\
&\quad \cup \cup_{p \in \Pi \wedge so(p)=[i_1, i_2, \ldots, i_n]} \{(i_{1R}, i_{2R}), (i_{2R}, i_{3R}), \ldots,\\
&\quad (i_{(n-1)R}, i_{nR}), (i_{nR}, v_p), (v_p, i_{1R})\};\\
d' &= \cup_{i \in V} \{((i_L, i_R), 0), ((i_R, i_R), 1)\}\\
&\quad \cup \cup_{(i,j) \in E_{int}} \{((i_R, j_R), d(i, j))\}\\
&\quad \cup \cup_{(i,j) \in E_{ext}} \{((i_R, j_L), d(i, j))\}\\
&\quad \cup \cup_{p \in \Pi \wedge so(p)=[i_1, i_2, \ldots, i_n]} \{((i_{1R}, i_{2R}), 0), ((i_{2R}, i_{3R}), 0), \ldots,\\
&\quad ((i_{(n-1)R}, i_{nR}), 0), ((i_n, v_p), 0), ((v_p, i_1), 1)\};\\
t' &= \{(i_L, lat(i)) \mid i \in V\} \cup \{(i_R, rat(i) \mid i \in V\} \cup \{(v_p, pro(p) \mid p \in \Pi\}
\end{aligned}
$$

where the functions lat, rat and pro are given, respectively, by

$$
lat(i) = \begin{cases} P(\pi(i)) - S(G, \pi(i)) & \text{if } sched(\pi(i)) = tdm \\ P(\pi(i)) - S(G, \pi(i)) & \text{if } sched(\pi(i)) = rr \\ 0 & \text{if } sched(\pi(i)) = off \end{cases}
$$

$$
rat(i) = \begin{cases} \frac{P(\pi(i)).t(i)}{S(G, \pi(i))} & \text{if } sched(\pi(i)) = tdm \\ t(i) & \text{if } sched(\pi(i)) = rr \\ t(i) & \text{if } sched(\pi(i)) = off \end{cases}
$$

$$
pro(p) = \begin{cases} 0 & \text{if } sched(\pi(i)) = tdm \\ P(\pi(i)) - S(G, \pi(i)) & \text{if } sched(\pi(i)) = rr \\ 0 & \text{if } sched(\pi(i)) = off \end{cases}
$$

where we simply used the formulas for the two terms of the latency rate models for TDM and NPNBRR schedulers.

5.3.11 Temporal Analysis Model for Partial Schedules

The construction of the temporal analysis model as given in the previous section assumes that the whole graph has been scheduled onto the target platform. During scheduling, we will need to check whether a partial schedule is feasible or not. In partial schedules, some actors may not yet have been mapped and the size of slices $S(p, G)$ allocated to the graph on a specific processor may not be yet known.

Our strategy when analyzing the temporal behavior of a partial schedule is to always take the most optimistic assumption about the overhead caused by the mapping of any actor that has not yet been yet assigned to a processor, with the objective of not excluding any feasible complete solution a priori. Sometimes, a particular combination of optimistic assumptions is clearly not feasible in practice. Say that actor B, which is yet unmapped, communicates both with actor A and actor C, which are respectively mapped to processor 1 and 2. Furthermore, A, B and C all map to the same processor type. We will assume that B communicates without inter-processor overhead with both A and C, although it is clear that B can be mapped to processor 1 or to processor 2, but not both. The reason for taking always optimistic assumptions for partial schedules, is that our scheduler, which we will present in detail in Sect. 5.5, searches for feasible schedules by assigning actors to processors one by one, and checking at each assignment whether the cumulative set of assignment decisions has not created an infeasibility. Taking pessimistic assumptions about unmapped actors could lead us to exclude feasible schedules from the search.

If the processor allocation of an actor $i \in V$ is known, and the slice size $S(\pi(i), G)$ has not yet been determined, then we make the slice size equal to the maximum slice $\hat{S}(\pi(i))$, since the final slice can never exceed that value, in the computation of the functions *rat* and *lat*.

As for an actor $i \in V$ yet unmapped, the $\pi(i)$ is undefined, which we will denote by $\pi(i) = \bot$. We must redefine the analysis model for this case. Edges where one of the endpoint actors is still unmapped are considered internal edges, if the processor type of the two is the same, and external if the processor type is different (independently of what mapping eventually is chosen, it is guaranteed that the two actors will not be in the same processor). Therefore, in generating a temporal analysis model for a partial schedule, the sets E_{int} and E_{ext} are given by

$$\begin{aligned} E_{int} &= \{(i, j) \mid (i, j) \in E \wedge (\pi(i) = \pi(j) \wedge \pi(i) \neq \bot)\} \\ &\quad \cup \{(i, j) \mid (i, j) \in E \wedge (\pi(i) = \bot \vee \pi(j) = \bot) \wedge \tau(i) = \tau(j)\}; \\ E_{ext} &= E - E_{int}. \end{aligned}$$

Furthermore, we must define the values of *lat* and *rat* for unmapped actors. In the best case, an actor $i \in V$ for which $\pi(i) = \bot$ will be mapped to the processor of type $\tau(i)$ that offers the highest maximum slice size, that is, in this case, we make $S(\pi(i), G) = \max_{p \in \Pi \wedge \tau(p) = \tau(i)} \hat{S}(p)$ in the equations for *lat* and *rat*. This is possible since we assume that all processors of the same type have the same dynamic scheduler settings (i.e. scheduler type and period).

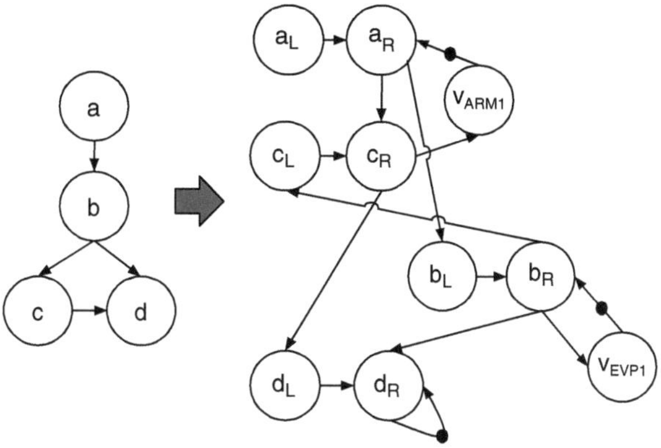

Fig. 5.5 Example of the generation of an analysis model for a partially scheduled graph

As an example, we depict in Fig. 5.5 a data flow graph representing the application (on the left) and its analysis model (on the right). The input graph has four actors a, b, c and d. The processor types to each this actors map are $\tau(a) = \tau(c) = ARM$ and $\tau(b) = \tau(d) = EVP$, that is a and c run on ARM processors, whereas b and d run on EVP processors. Furthermore, we are mapping this application to a virtual platform that includes several ARMs and EVPs. One of the ARMs is named ARM1 and one of the EVPs is EVP1. Assume that the current partial schedule includes only mappings to these two processors, and it is the following: the static order for ARM1 is $so(ARM1) = [a, c]$, and, for EVP1, is $so(EVP1) = [b]$. Since each processor runs a scheduler (for the topology it does not matter whether it is TDM or NPNBRR), we insert a latency-rate model for each actor. We add edges (a_R, c_R), and edges (c_R, v_{ARM1}) and (v_{ARM1}, a), to represent the static order. The processor actor is required to accurately model the budget replenishment interval for NPNBRR, and its execution times is 0 for a TDM scheduler. Application edges between actors mapped to different processors (such as (a, b)), or between actors mapped to different processor types (such as (c, d) are represented in the model by edges between the R actor of the producer and the L actor of the consumer. An application edge between actors mapped to the same processor, or, alternatively, having the same processor type, where at least one of the actors has not been mapped yet (such as (b, d)), are represented in the model by an edge from the R actor of the producer to the R actor of the consumer, since it may still be possible for the two to be mapped to the same processor.

5.4 Combining CTS and RTS

Although static order scheduling is a popular strategy for scheduling data flow graphs [81], it cannot singly solve the mapping problem we proposed in Chap. 2. Since jobs start and stop independently, a static order schedule would have to be computed at design time for every combination of jobs that can be active

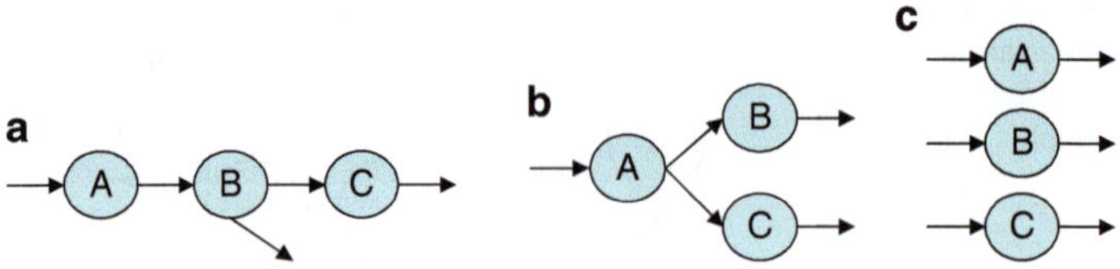

Fig. 5.6 Three job fragments

simultaneously. This problem is aggravated by the fact that, when a transition occurs from a job-mix to another, remaining jobs must not experience any discontinuity caused by the change of configuration. Secondly, as different jobs can have very different rates, a static schedule could only be realized if the faster jobs where almost completely latency-insensitive, which is certainly not the case with jobs such as Wireless LAN and, even if this were the case, the length of the static order schedule and the amount of buffering of inputs required could easily become prohibitive.

One could think of simply using local TDM or NPNBRR schedulers per processor and relying on FIFO communication for actor synchronization. The main problem with such a strategy is that the bounds on the worst-case response times of actors executing on independent slices completely overlook the fact that, within a job, we have more information about the interdependence of actors. For instances, a set of actors may be mutually exclusive – in an SRDF graph this happens when the actors belong to the same single-delay cycle – and allocating a different slice to each of these wastes resources, since, if all share the same slice, each can use the whole slice when enabled.

To illustrate this, we compute the worst-case latency from input to output for three groups of three actors, where all three actors belong to the same job and are allocated to the same processor. These are shown in Fig. 5.6. In two of the cases, there are direct dependencies amongst them. In the third case no such dependencies exist. Edges without a source receive their input tokens from actors executing on other processors, which are not depicted.

Assume that each actor has an execution time of $t = 1$. Figure 5.7 shows two different schedules, both valid for all of the three cases, assuming a TDM wheel period $P = 4$. In schedule Fig. 5.7a each actor gets its own time slice of $S = 1$ duration. In schedule Fig. 5.7b, the actors are statically ordered, and a slice of $S = 3$ duration is allocated to the statically scheduled group. Both schedules require exactly the same amount of processing resources (3/4 of the time wheel), but worst-case response times are much smaller for the combined scheduling strategy.

If the actors in Fig. 5.6a are scheduled according to the schedule in Fig. 5.7a, the output of B will be produced, in the worst-case, after A and B have both executed once. A will take $r(A) = 4$ to execute, because in the worst-case A has to wait for 3/4 of the wheel to turn to get its slice, and then take $1/4$ of the wheel to process. Now B has to execute once. That takes $r(B) = 4-1$, since B can execute sooner that waiting for a full turn after A has completed – we know that B is not allocated to the same slice as A. The output of B will therefore be generated after $r(A) + r(B) = 7$

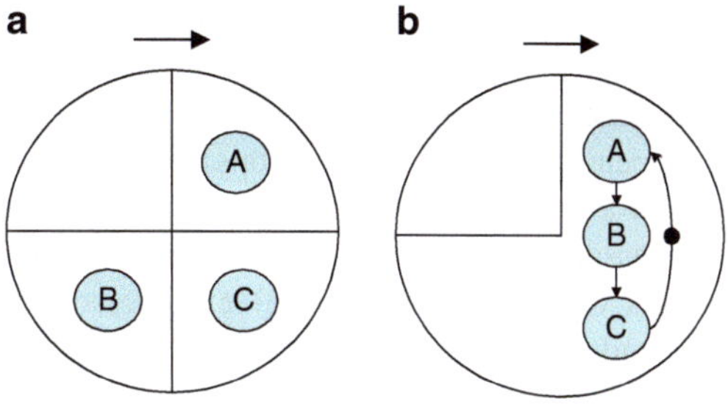

Fig. 5.7 Two schedules of three actors. (**a**) TDM. (**b**) Combined

Table 5.1 TDM vs combined scheduling

	TDM			Comb.
	Job a	Job b	Job c	All
Out A	4	4	4	2
Out B	7	7	4	3
Out C	10	7	4	4

units of time. The output of C will still have to wait for the response time of C, which will be again less than one period, since the slice allocated to C will certainly be reached before the slice of B, which just finished. Therefore $r(C) = 3$, and C will produce output after $r(A) + r(B) + r(C) = 10$ time units. On the other hand, the combined TDM/static order schedule Fig. 5.7b takes as worst-case the time from the end of the slice allocated to the group to the next end of the slice allocated to the group, since the three actors all execute in sequence within one slice. Therefore, A ends execution at time 2, the output in B is produced at time 3 and the output of C has a response time of 4.

For job Fig. 5.6b, the TDM schedule in Fig. 5.7a will yield a response time of $4 + 3 = 7$ for both external outputs, while the combined schedule will yield the same values as for job Fig. 5.6a.

For job Fig. 5.6c, the combined schedule imposes an arbitrary order between three independent actors, and thus creates extra scheduling dependencies. However, the worst-case production times, assuming all external inputs are available, are still 2, 3, 4, for the outputs of A, B and C, respectively, while for the TDM schedule, the worst-case production times are 4, 4, 4. Note however that in this case the combined schedule imposes an extra dependency on the arrival of the input of A to the production of B's output, which did not exist before. If the input of B is ready before the input of A, this particular schedule may be quite inefficient. However, this is a problem of choosing the "right" static order. The results of this example are summarized in Table 5.1.

The combination of TDM with static order schedule allows, when applicable, the determination of much tighter bounds on worst-case response times for the same amount of allocated resources than TDM. A similar example could be given for NPNBRR schedulers. Since at compile time we know only about one single job, it is at this level that static order can be used to shorten worst-case bounds on response times.

Because of this, we use different scheduling methods to schedule among tasks within a job (intra-job scheduling) and to reserve resources among jobs (inter-job scheduling) such that each running job can meet its timing requirements independently of any starts/stops of other jobs.

Intra-job scheduling is handled by means of static order, i.e., per job and per processor, a static ordering of actors is found that respects the Real-Time requirements while trying to minimize processor reservation, while inter-job scheduling is handled by means of local TDM/NPNBRR schedulers. In the case of TDM, scheduling, per job and per cluster of statically-ordered actors, a slice time with duration S must be determined.

An admission controller is needed to start jobs upon the arrival of a start request. It is similar to the one we suggested in [73], in that it must check if enough memory, communication and computing resources (enough time in the TDM wheel of the processor) are available for the requirements of the job. If not, the start of the job is refused. The mapping algorithms used by the admission controller are described in detail in Chap. 7.

5.5 The Scheduling Problem

Our objective is to schedule each job in such a way that we guarantee real-time constraints and minimize processing resources reservation, such that resources can be shared with other running jobs, both increasing the probability of starting the job on an already running system, and the probability that jobs requested to start later will find enough resources to start. We need to define the period $P(p)$ of the time wheel on each processor $p \in \Pi$, the slice time $S(p, G)$ attributed to job G on processor p, a static order schedule $so(p, G)$ per processor p and job G and sizes for all buffers. The maximum acceptable period between consecutive executions of an SRDF actor for a given job graph G is given by μ_D.

In order to propose a solution for this problem, we must first discuss optimization criteria. After this, we will discuss how to decouple different decisions in the scheduler, and we will then describe our proposed solutions for each phase.

5.5.1 Optimization Criteria

For each job, the schedule should reserve as few cycles per period as possible, as this will increase the percentage of the processor available to other jobs. A good measure of the amount of processing resources reserved by a job G on a processor p is the processor reservation $U(p, G)$, which can be defined, for TDM, as the ratio between the time reserved on the processor for the job and the total number of cycles per time wheel period:

$$U(p,G) = \frac{S(p,G)}{P(p)}. \tag{5.9}$$

The processor period $P(p)$ is, as we have seen, defined per processor as a fixed system parameter which is considered an input by the scheduler. We will discuss in Sect. 5.5.3 how to appropriately chose the value of this parameter.

For a heterogeneous multiprocessor, we try to reduce the reservation on all processors. This is done in a weighted way, since it is possible that certain processors types are in general more required than others, and thus their reservation should be kept lower. We introduce a processor cost coefficient $c(p)$.

Our optimization criterion is the minimization of the weighted sum of the reservations:

$$\text{minimize} \sum_{p \in \Pi} U(p,G) \cdot c(p) \tag{5.10}$$

This function still disregards the fact that, although the resources of a processor may be scarcer than of another, if the "cheaper" processor is fully occupied, several jobs may become impossible to start – e.g. a second instance of an already running job. This could be accounted for by an optimization criterion that makes the cost of a processor increase with reservation, but that would yield a complex, non-linear objective function. Instead, we rely on defining our maximum slice size per job, $\hat{S}(p)$ per processor as part of the input to the scheduler, and enforcing that $S(p,G)$ must be kept lower than $\hat{S}(p)$ for any valid schedule.

Another potentially important optimization criterion is the amount of buffering required in total and by processor. In our work we decided to assume that buffering was less important than processor reservation. This is because in the radio models that we studied buffer sizes were typically small and mostly independent from the other resource allocation decisions, as we shall see in the results section.

5.5.2 Phase Decoupling

The problem of trying to find $P(p)$, $S(p,G)$, $so(p,G)$, and $b(i,j)$, subject to the afore-mentioned constraints and optimization criterion is complex. It is therefore necessary, when designing the scheduler, to try to define a number of phases in which each of the decisions is taken.

The choice of the periods of the schedulers, $P(p)$, should be independent of the actual jobs, and these values should simply be given as inputs to the scheduler. Nonetheless, the choice of periods is a very important decision, and one that can be made much easier by taking into account the characteristics of the transceivers that one knows will be required to run in the platform. Section 5.5.3 discusses how to chose these values.

For NPNBRR schedulers, the choice of $S(p, G)$ is made along the choice of $so(P, G)$. This is because, for a NPNBRR scheduler, since preemption is not possible, the slice sizes are simply equal to the execution times of the tasks allocated to the processor. For TDM schedulers, we have an extra phase of the compiler dedicated to trying to reduce the size of the slices, after a feasible $so(P, G)$ has been found. A third phase takes care of computing feasible buffer sizes for all FIFOs, and can be applied before, after or in-between the two other phases, depending on how many buffering resources are available in the platform.

5.5.3 Determining Scheduler Settings

For an NPNBRR scheduler, we do not have many options in the choice of the period $P(p)$ of processors. Say that we want to be able to run a number of different jobs, each with a different μ_D. Since for the lowest μ_D it is still required that each actor fires once per period, we should set $P(p)$ at most to the value of the lowest μ_D across all the jobs we intend to map in the platform. This value is likely to make other jobs unschedulable, if they are composed of actors, with execution times too high to fit in a single turn of the Round Robin wheel. In such a case, one can consider that when such a job is running, there can be no sharing of resources with the faster job, and the compile time scheduling can be carried out with a higher μ_D. This will, however, make it impossible for this job to share resources with any job with a faster μ_D for that processor type. Also, since one of our given requirements (see Chap. 2), is that we cannot know at compile time all other transceivers that may be mapped in the platform, our choice of $P(p)$ can only be based on the jobs that are known at design time, and on known characteristics (such as sampling rates) of transceivers that may in the future be added to our platform.

For TDM schedulers, assuming a constant processor reservation, we notice by inspecting Eq. 5.4 that when the time wheel period decreases, the worst-case response time of an actor becomes closer to its execution time. This suggests that the period of the TDM scheduler of each processor should be as small as possible, while still allowing partitioning of processing resources among the maximum amount of job instances one wishes the platform to be able to run simultaneously. However, this neglects the overhead caused by the context-switches that occur when changing slices: since the context-switching time is constant, a smaller time wheel period implies a higher absolute number of context switches and, thus, less time left to be used for reservations on the processors. We can chose $P(p)$ in such a way that we guarantee that the overhead of context-switching does not exceed an arbitrary percentage $pcc(p)$ of the processing cycles of p. If $cc(p)$ is the cost of a context-switch for processor p and $nj(p)$ is the maximum number of job instances we wish to run simultaneously, and since per each period each job gets one time slice, then the number of cycles per period spent on context switching is $cc(p) \cdot nj(p)$ and the value of $P(p)$ should be set such that:

Fig. 5.8 A job

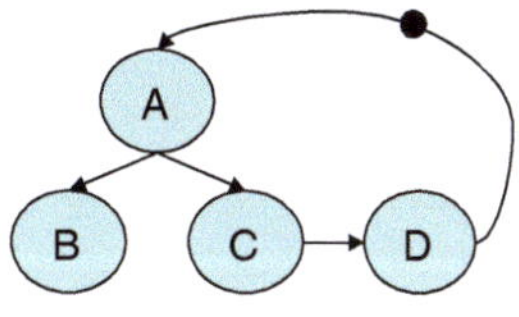

$$P(p) \geq \frac{cc(p) \cdot nj(p)}{pcc(p)} \tag{5.11}$$

On the other hand, since any actor in a timed SRDF graph must be able to execute once per μ_D of its job to meet the temporal constraints, the time wheel period must be such that allows this for all jobs, and therefore $P(p) \leq \mu_D(G)$ for any job G that the system must support. If the two inequalities are not compatible, one must accept a higher percentage of context-switching overhead.

5.5.4 Finding Static Order Schedules

Our data flow scheduler groups actors into clusters, while simultaneously ordering actors statically within each cluster.

The first thing to notice about the static order schedules is that the only precedence constraints that need to be respected are the ones inside one iteration. This is because the static order imposes an execution order with no iteration overlap per cluster (but notice that between two different clusters there may still be iteration overlap). Therefore, for the purpose of the static order schedule, only edges in the input graph that have the delay equal to 0 need to be taken into account.

To give an idea of how the static order of actors on a processor influences the temporal behavior, we will give an example. Figure 5.8 depicts the timed SRDF model of a job. Assume that actors A, B and C are allocated to a cluster and D to another. Assume also that $t(A) = t(B) = t(C) = t(D) = 1$ and that $\mu_D = 3$. The MCM of the job is $\mu = 3$, because of the ACD cycle.

Now inspect the two clustered static order schedules in Fig. 5.9. Both of them respect the precedence constraints defined by the 0-delay edges, but while the schedule depicted in Fig. 5.9a has an MCM of 3, from cycles ACD and ACB, the schedule depicted in Fig. 5.9b exceeds μ_D since the cycle $ABCD$ has $\mu_c = 4$, due to the extra dependency caused by the static order on the first cluster.

The data flow scheduler works thus: first, a Direct Acyclic Graph (DAG) representing the dependencies between actors within one iteration is generated, by removing all edges with delays from the input graph. Then the scheduler starts. It picks an actor without dependencies. It generates a list of processors to which the selected actor can be mapped. It then selects one of these processors, and tries to map the selected actor to the selected processor. This appends the actor to the statically-ordered sequence of a given processor.

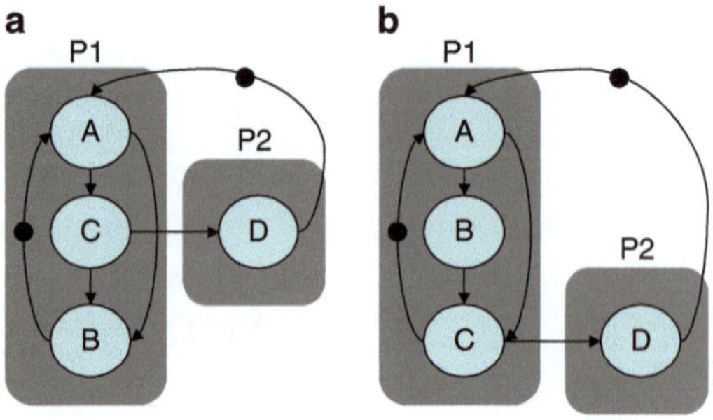

Fig. 5.9 Two schedules for the same job

After each mapping attempt, a temporal analysis model for the obtained partial schedule is constructed, and an MCM computation checks whether the MCM of the temporal analysis model is lower than the required μ_D. For MCM computation, we implemented the Howard algorithm [13], since it is one of the fastest, according to the benchmark provided in [17] and, being a policy-improvement algorithm, it is easy to write an incremental version of it: at each step we can start the search of the MCM of the temporal analysis model by the MCM cycle which was the solution in the previous step (assuming that cycle still exists in the new temporal analysis model).

If the MCM of the temporal analysis model is below μ_D. The last scheduling decision is kept, the DAG is updated by removing the actor that was mapped, a new actor is considered from the fireable nodes in the DAG. If the mapping fails to meet the timing requirements, the scheduler undoes the last actor to processor mapping and tries another of the possible processor mappings for that actor. If no possible allocations are left for this actor, it backtracks, undoing its previous last successful actor to processor mapping, and trying another one.

The algorithm finds a solution if it runs out of actors to allocate, and all actors are allocated to a processor. It fails if no valid solution can be found after all sequences of mappings have been tried.

This algorithm guarantees that actors are presented to the scheduler in all possible orders that respect the intra-iteration dependencies (as given by the DAG), and that all processor mappings for every ordering are considered. The problem is that the same static order schedule can be tried several times. For instances, reversing the order of mapping two actors with different processor types will result in the exact same static order schedule. To avoid this, we can keep a hash table with all the static orders that we have encountered before, and backtrack every time we find ourselves in a previously explored state.

5.5.5 Unrolling and Static Ordering

A limitation of the static order scheduler as given is that it does not allow interweaving the execution of several iterations of the graph within the static order of one processor.

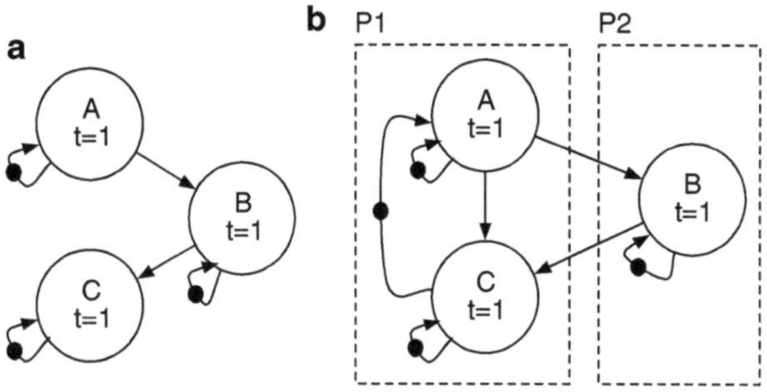

Fig. 5.10 An input graph and the timing analysis model resulting from statically ordering this graph in two processors

An example is given in Fig. 5.10a. The graph has three actors A, B, and C, where A and C must be scheduled onto processor P1 and B must be scheduled onto processor P2. Since the scheduler considers only one iteration at a time, it can only find the static order given in Fig. 5.10b.

We can allow iterations to be inter-weaved within a static order by considering several graph iterations at a time. This can be done by unrolling the graph. Unrolling an SRDF graph allows us to represent in a single iteration several iterations of a SRDF graph – it is essentially equivalent to treating the input SRDF graph as an MRDF graph, setting the repetitions of each actor to the desired *unroll factor*, and performing MRDF to SRDF conversion as explained in Chap. 3. The result of unrolling the SRDF graph given in Fig. 5.10a by a factor of 2 is depicted in Fig. 5.11a. Each of the original actors is now represented by two copies. If given as an input this unrolled graph, the scheduler can now find a scheduler with a higher throughput, such as the one depicted in Fig. 5.11b. In this new schedule, because we allow two copies of A to be scheduled one after the other, we obtain an analysis graph that has an MCM of 4. Since this is a graph that represents two iterations of the original graph, we get two firings of each original actor for every four time units, an effective throughput higher than the throughput of the schedule in Fig. 5.10b, where due to the ABC cycle caused by the static ordering we get an MCM of 3, resulting in one firing of each actor for every three time units.

Pseudo-code for the Scheduler

In what follows, we will present the scheduler functions in a pseudo-code that is mostly composed of valid Objective Caml functions, except for the statements starting by "for each" which we use as an easier to read placeholder for what in the Objective Caml implementation is a call to `List.iter`. The top-level `schedule` function works as the external call point for the scheduler. Its input argument is a graph. A graph is an ordered pair of nodes and edges. The function extracts the initial DAG from the graph and initializes an empty list of binds, after which it makes the initial call of the `scheduling_step` function. The call will always return an exception that

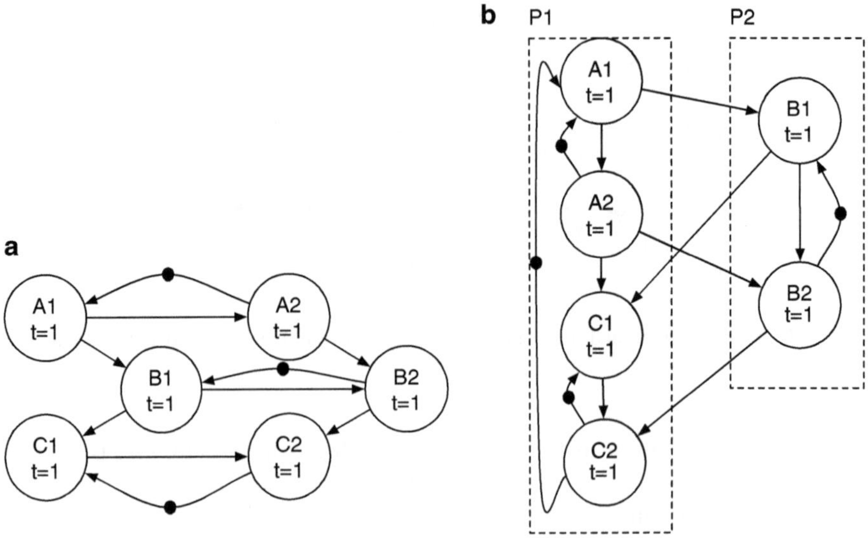

Fig. 5.11 The input graph of Fig. 5.10a unrolled by a factor of 2 and the resulting timing analysis model for a possible static-order schedule of this graph in two processors

can be caught by the caller. If the call to scheduling_step succeeds, it will throw the exception Solution_found s, where s is the complete valid schedule, if not, schedule will throw the exception No_Solution_Found.

```
let schedule graph=
  (*scheduling dag is the graph without edges with delays*)
  let dnodes= nodes graph
  and dedges= filter (fun e -> delay e < 1) (edges graph) in
  let dag= (dnodes, dedges)
  and binds= [] in
  (*a schedule is the pairing of a dag with a list of binds*)
  let empty_schedule= (dag, binds) in
  scheduling_step empty_schedule;
  raise No_Solution_Found
```

The recursive function scheduling_step goes through all the actors that are ready to be mapped in the current partial schedule and tries to map each to each processor where it can run. If a mapping does not infringe the temporal requirements, which is checked by calling eval_schedule, scheduling_step invokes itself to continue the mapping procedure. If the mapping fails, the next (actor, processor) pair is evaluated. If a complete valid schedule is found, it raises the exception Solution_found s, where s is the complete valid schedule. If a mapping is invalid, the call to scheduling_step will simply return. Note that we left outside of this description the hashing of the static order schedule that allows us to recognize a scheduling branch that has been previously evaluated. This was omitted to make the program easier to read.

```
let rec scheduling_step schedule =
```

```
let (dag, binds) = schedule in
(*an actor is fireable if it has no predecessors in dag*)
let fireable_actors = get_sources dag in
if dag = ([],[]) then raise (Solution_Found schedule);
for each actor in fireable_actors do
  for each processor in (possible_mappings actor) do
    let new_schedule = map actor processor schedule in
    let eval_outcome = eval_schedule new_schedule graph in
    match eval_outcome with
    |Valid -> scheduling_step new_schedule
    |Invalid -> ()
   done
done
```

The eval_schedule function generates an analysis model for the given schedule and graph and evaluates wether the analysis model meets the temporal requirements, returning either Valid or Invalid. The possible_mappings function returns a list of processors in the multiprocessor architecture that match the type of processor of the actor in the function argument. A scheduling heuristic can be defined by reordering the fireable_actors list, the possible_mappings list, or both. For instances, ordering the fireable_actors list such that the actor with the longest path to the sinks of the DAG is mapped first can significantly improve the number of steps the scheduler takes to find a solution, at least for some graphs. The map function adds the new actor to processor mapping to the bind list. Binds are represented as (actor, processor) pairs.

```
let map actor processor (old_dag, old_binds) =
  let new_binds = (actor, processor) :: old_binds
  and new_dag = remove_node old_dag actor in
  (new_dag, new_binds)
```

The static order on a processor can be extracted from the binds list by going through the binds list and obtaining a list (in reverse order, since the scheduler always adds the last bind at the head of the list) of all actors in pairs (a,p) where p = proc.

```
let so proc binds =
  fold_left (fun l (a,p)-> if p=proc then a::l else l) binds
```

5.5.6 *Finding the Slice Times*

For a TDM scheduler, we can try to reduce the size of the slices allocated to a given job on a given processor. This is not possible for NPNBRR, because of the lack of preemption – the size of the slice in this case matches the sum of the execution times of tasks mapped to the processor.

Convex Programming Formulation

If we write all the SPS precedence constraints (Eq. 4.5) for the temporal analysis graph, as a function of the start times and the slice sizes, we obtain a set of constraints that limits our choice of the slice sizes at each processor, while meeting the desired period μ_d. If we couple this with an objective function that minimizes the sum of slices, we obtain something that resembles the structure of a linear program. Given the static order per processor $so(p)$ (we drop the job parameter G, since when solving the slice time problem this can be kept implicit, and we do the same for slice size S) and actors i and j belonging to the input graph, let $sc(so, i) = j$ if j is the actor after i in one of the static orders in so. Also, let $tl(so, p)$ provide the last actor in the static order of processor p, and $hd(so, p)$ provide the first element in the static order for processor p. We then write a constraint requiring every start time to be greater or equal to zero, every slice to be greater or equal to zero, and lower than the maximum slice size of the processor, and an equation derived from Eq. 4.5 for each edge in the analysis model. Then the following non-linear program accurately represents all the sequence constraints in the temporal analysis graph, and the execution times of all actors in the analysis model:

Minimize	$\sum_{\forall p \in \Pi} S(p)$
subject to	
$\forall i \in V$	$s(i_R) \geq 0$
$\forall i \in V$	$s(i_L) \geq 0$
$\forall p \in \Pi$	$s(v_p) \geq 0$
$\forall p \in \Pi$	$S(p) \geq 0$
$\forall p \in \Pi$	$S(p) \leq \hat{S}(p)$
$\forall (i, j) \in E_{int}$	$s(j_R) - s(i_R) - \frac{t(i).P(\pi(i))}{S(\pi(i))} \geq -\mu_d * d(i, j)$
$\forall (i, j) \in E_{ext}$	$s(j_L) - s(i_R) - \frac{t(i).P(\pi(i))}{S(\pi(i))} \geq -\mu_d * d(i, j)$
$\forall i \in V$	$s(i_R) - s(i_L) + S(\pi(i)) \geq P(\pi(i))$
$\forall i, j \in V : sc(so, i) = j$	$s(j_R) - s(i_R) - \frac{t(i).P(\pi(i))}{S(\pi(i))} \geq 0$
$\forall p \in \Pi : i = hd(so, p)$ and $j = tl(so, p)$	$s(i_R) - s(j_R) - \frac{t(j).P(p)}{S(p)} \geq -\mu_d$

The objective function minimizes the sum of TDM slices. An arbitrary weight can be multiplied by each $S(p)$, in the objective function, such that we introduce different costs for using processing resources on different virtual processors. The variables of this formulation are the slice times $S(p)$ and the start times $s(i)$.

The constraint set is non-linear, as the $S(p)$ variables appear inverted in some of the constraints. Therefore, a linear programming solver cannot handle this problem. We can however, use the results of the work in disciplined convex programming [34], which defines a set of rules for the constraints and objective function that, if followed, allow the non-linear program to be classified as belonging to a subset of non-linear programs, so-called convex programs, because they have a convex solution space, and solved by a convex programming solver, which can be implemented with polynomial complexity on the size of the technology matrix (i.e., on the number of variables and constraints).

Our problem is in fact a Disciplined Convex Program (DCP). We will proceed to show that this is the case according to the requirements of DCP as stated in [34]. First, for the program to be a valid DCP, the objective function must either be the minimization of a convex expression or the maximization of a concave expression. According to DCP rules, a sum of convex expressions is a convex expression; a variable multiplied by a constant is an affine expression, and affine expressions are both convex and concave. Our objective function is the minimization of a sum of affine expressions, and therefore a valid DCP objective function.

As for the constraints, DCP rules prescribe that the constraints must have the structure where a convex expression is lesser or equal than a concave expression.

We've already seen that variables multiplied by constants are affine, and may be interpreted as either concave and convex. A DCP construction rule specifies that the sum of affine expressions is also affine. So many of our constraints are of the type affine expression is greater or equal to affine expression, which are valid DCP constraints $(concave \geq convex)$.

Our only problem is the expression $-\frac{t(i).P(\pi(i))}{S(\pi(i))}$ where $S(\pi(i))$, a variable, is inverted. According to DCP rules, inversion is a function that is neither convex nor concave. More specifically, it is convex for positive values of the variable, and concave for negative values of the variable. In our case, however, our variable $S(p)$ must always assume positive values, since a negative slice has no meaning. CVX, a DCP solver [33], allows us to use a function called `pos_inv` to indicate an inverted variable that can only assume positive values. Thus all our occurrences of $\frac{1}{S(\pi(i))}$ can be recognized by CVX as being convex. A DCP rule states that a convex expression multiplied by a negative constant results in a concave expression. Thus all our left hand sides can be construed as concave expressions (by being sums of affine and concave expressions) and all our right hand sides are affine, and therefore convex. Since constraints of the type $(concave \geq convex)$ are valid DCP constraints [34], we have shown that all our constraints are valid DCP constraints, and our problem can be solved with polynomial complexity on the size of the technology matrix by DCP – and therefore on the number of actors in the graph, for a fixed number of processors in the virtual platform.

Our convex program will provide a solution that is optimal for the weights we provided for the processor slices, in polynomial time.

Linear Programming Formulation

If a DCP solver is not available, we can derive a Linear Program that provides a conservative approximation. We can linearize the problem as follows. First, we remove all the non-inverted occurrences of $S(p)$ from the constraints in our convex program – we do this because, looking at our analysis model – we are simply overestimating the latency in the latency-rate model from $P(p) - S(p)$

to $P(p)$, so the result is conservative. Furthermore, the smaller the slices are, the smaller the overestimation will be. Then, we make a variable substitution: $N(p) = P(p)/S(p,G)$. Instead of maximizing the total amount of used time slack, we maximize the sum of $N(i)$, weighted by the processor costs as defined in Sect. 5.5.1. Since all expressions in both the objective function and constraints are affine, the problem is a linear program.

A problem with this linear programming approach is that the approximation of $P - S$ by P may incur in very conservative results in cases where latency is the dominant factor. We will now present two different ways of optimizing the slice sides that can be used to further reduce the size of slices without requiring the use of a DCP solver, or the approximation and a linear programming solver.

Binary Search Slice Allocator

This algorithm performs a binary search on each of the scheduled clusters of actors to try to allocated as much of the deadline extension to it as possible. Clusters are sorted according to the weight of the processor where they are mapped. The maximum deadline extension is calculated per group, using an all-pairs shortest path algorithm. Starting with the group with the highest priority, we calculate the new slice time using a binary search between maximum and minimum slice values. A minimum slice value can be inferred from the minimum deadline extension pool across all actors mapped to the same virtual processor. In pseudo-code, the code of the binary search slice allocator is:

```
let bin_search (graph, min_slice, max_slice, last_slice) =
  if max_slice <= min_slice then last_slice
  else
    let new_slice = (max_slice + min_slice)/2 in
    let new_mcm = compute_mcm (graph, new_slice) in
    if new_mcm > max_mcm then
       bin_search (graph, new_slice+1, max_slice, last_slice)
    else
      bin_search (graph, min_slice, new_slice-1, new_slice)
```

The algorithm can be sped up by not using the MCM computation algorithm, but instead relying on using a negative cycle detection algorithm such as Szymanski [17] to check if the current MCM is still below the maximum MCM allowed (i.e. μ_D).

The results for this algorithm, however, will depend heavily on which processor and group we start the binary search. The algorithm lowers as much as possible the slice time for the first cluster it performs the binary search on, sometimes leaving little or no slack to allow the remaining clusters to lower their slice times. In fact, its performance is in practice very similar to the linear program, with the advantage that it does not use an approximation for the execution time of the latency actor.

This allocator applies cycle detection at every iteration of a binary search. It has polynomial complexity since both cycle detection and binary search have polynomial complexity.

Randomized Slice Allocator

We observed that the binary slice allocator depends quite heavily on the clustering ordering on which the search is performed. To overcome this potential weakness the randomized slice allocator selects randomly clusters to decrease their slice time by a specific amount, thereby trying to be spread budget by all clusters. As in the previous algorithm, if the scheduled graph respects the specified throughput for some predefined slice times, this algorithm will find smaller or equal slice times for the graph thereby decreasing the throughput to as close as possible to the specified minimum throughput.

The random slice allocator adds all clusters to an open list (i.e. a list of clusters where we may still decrease the slice) and select a cluster from there. The probability of a cluster being chosen for slice decrease is dependent on its weight. That cluster sees its slice time decreased by a specified amount (pre-defined step of the algorithm) and the graph is tested for compliance with the throughput constraints. Should the smaller slice result in a violation of imposed constraints, that cluster is removed from the open list and its slice value restored to the previous amount. This select and decrease step is repeated until there are no clusters in the open list, which means it is not possible anymore to lower slice times while respecting the throughput constraints.

This algorithm depends also on a cycle detector. Essentially it runs the cycle detector a number of times which depends on the step size and on initial cluster's slice times. The maximum running time for this algorithm can be bounded and is also a class P algorithm.

5.5.7 Buffer Sizing

Ning and Gao [76] propose an approach to compute minimal buffer sizes for an SRDF graph G to allow for rate-optimal static-periodic execution of the graph, that is, the graph can execute a static-periodic schedule with period equal to its MCM. Ning and Gao employ a linear program formulation for this problem which they claimed to be optimal, implying that the problem is solvable in polynomial time, since linear programming is solvable in polynomial time.

In [72], it has been shown that Ning and Gao's formulation is not optimal, and that the problem of finding a ROSPS for an arbitrary SRDF graph that results in minimal total buffer size is, in fact, NP-complete. Furthermore, this paper shows that the constraints of the linear program used by Ning and Gao is incorrect and proposes a corrected version.

Another interesting result in [72] is that the set of optimized buffer sizes computed for a ROSPS of the graph of interest are valid for any self-timed execution of the same graph with equal or lower execution times.

This means that sufficient buffer sizes for our self-timed implementation can be computed by using the linear programming formulation originally described in [76] and corrected in [72]. The technique is based on the observation that the maximum amount of tokens $b(i, j)$ that needs to be stored in the FIFO corresponding to arc $(i, j) \in G$ between the start of any firing of the producer i and the end of the correspondent firing of the consumer j, is lower-bounded by the number of firings of i that occur in that interval. The length of the interval is given by $s(j, k + d(i, k)) - s(i, k)$ and since in a ROSPS the firings of i occur at a periodic spacing of μ_D time units, the bound on the buffer size is given by:

$$b(i, j) \geq \frac{s(j, k + d(i, j)) + t(j) - s(i, k)}{\mu_D}. \tag{5.12}$$

Please not that since we have an SPS schedule, we can relate all start times of firings of an actor to its first firing, as explained in Chap. 4, since $s(i, k) = s(i, 0) + k \cdot \mu_D$, for all $i \in V(G)$. Thus an SPS schedule of G that minimizes all such distances can be constructed by taking the linear program we presented in Chap. 4 for the construction of a ROSPS of G and adding a constraint for each edge (i, j) of the following form:

$$b(i, j) \geq \frac{s_{SPS}(j) + t(j) - s_{SPS}(i) + \mu_D.d(i, j)}{\mu_D}. \tag{5.13}$$

We also must replace the objective function by one that optimizes the sum of all buffer sizes weighted by the size of the tokens that are produced in that arc:

$$\text{maximize} \sum_{(i,j) \in G} b(i, j).z(i, j) \tag{5.14}$$

where $z(i, j)$ is the size of each token produced in arc (i, j).

As for the expected performance of this approximation, it is shown in [72] that, for a randomized input data-set, the difference between this linear program and the optimal solution appears to be in many cases negligible. For a randomized set of connected graphs with 40 nodes, only in 50 % of the cases does the linear program overestimated the buffer sizes, and for these, the average overestimation was 2.1 %. The results also seemed to indicate that with the increase of the size of the graphs, the likelihood of overestimation would increase, but the size of the overestimation would decrease.

In our toolset, we employ the corrected version [72] of the linear program. Other approaches can also be used, such as the one presented in [86], which can yield optimal buffer sizes but suffers from exponential complexity, or the approach taken in [90], that applies heuristics to minimize the buffer size for CSDF graphs.

5.5.8 Scheduling Multi-rate Graphs

A Multi-Rate Data Flow graph (or, for that matter, a Cyclo-Static Data Flow graph) can be scheduled using the framework described in this book, by converting the original MRDF graph to its equivalent SRDF graph. The conversion does not have polynomial complexity – the number of copies per actor in the MRDF/CSDF graph is given by the number of repetitions in the repetition vector (see Chap. 3).

Care must be taken to ensure that when a SRDF copy a_0 of a MRDF actor a is mapped to a virtual processor $p \in \Pi$ during the static ordering phase, all other SRDF copies a_i of the same MRDF node are allocated to the same processor. The analysis model is also changed to reflect this fact: the values of lat and nat functions for all a_i must be computed assuming an allocation to p. If, due to backtracking, the decision to map a_0 to p is undone, the allocation of all copies must also be undone.

No changes are necessary to the slicing algorithms.

As for buffer sizing, a separate buffer $b(i_k, j_l)$ can still be calculated for each arc $(i_k, j_l) \in E$, where i_k is the copy number k of an MRDF actor i, and j_l is the copy l of an MRDF actor. Since in the implementation a single buffer is shared to connect all copies of i to j, the static-periodic schedule resulting from the buffer sizing linear program is no longer buffer optimal. However, if we simulate this static periodic schedule, and count the maximum number of tokens that are accumulated in every edge during the simulation, we obtain a feasible buffer size for each one of the buffers in the MRDF graph. Govindarajan et al. propose another linear-programming approach [31] that obtains better results by directly deriving buffer constrains from the MRDF graph. According to the authors, this approach can obtain, in some cases buffer sizes that are about half of the buffers sizes obtained by the approach we sketched above. The problem with this approach however, is that the number of constraints in the linear program for each edge (i, j) is given by the product of the repetitions vector entries $r(i)$ and $r(j)$, and this means that the linear program can in some cases become too large for the solver we used in our implementation – the GNU Linear Programming Kit [26] – to handle. Our toolset implements both approaches and leaves to the user the choice on which to use. Only one of the transceiver graphs we used in our experiments is a MRDF graph, namely the WLAN receiver presented in Sect. 5.6. In this case, the two approaches delivered the same results.

One possibility that arises from the fact that we can formulate both the slicing problem and the buffer sizing problems as convex programming problems with a conservative linear programming approximation, is that the two can be combined onto a single convex programming problem that solves them in tandem. In [91], the authors propose exactly such a formulation. However, the authors did not realize that the problem can be solved exactly (for time expressed as a real number) by a convex solver (at least by CVX), and instead use an approximation to get rid of the inverted appearances of the slice variables (as we have shown, inversion is a convex function for variables that are guaranteed to be positive).

5.5.9 Phase Ordering

It is not trivial to chose in which order to perform the determination of the slice times and the computation of the static order schedule, since these two steps are strongly interdependent. If one determines the slice times first, the determination is based on partial constraints since the schedule is yet to be derived. It may be that there is no valid static order schedule that meets the tighter scheduling constraints caused by replacing the original execution times by the larger response times due to the chosen time slicing.

An important consideration is that all the algorithms we propose are polynomial, except for the static order scheduler, which has exponential complexity, and the MRDF to SRDF expansion, if the starting graph is not single-rate.

We can determine the static order schedule first, using response times based on the maximum allowed slice per processor, $\hat{S}(p)$, and only afterwards try to find slice times that are compatible with that static order schedule. Thus, we first approach the problem as a constraint satisfaction problem, to which the static order scheduler delivers a feasible solution, if there is one, and then try to optimize this solution by reducing utilization. This has the advantage that, provided there is a feasible static order, there is always a solution. However, it may be that the static order schedule chosen is not optimal in the sense that it does not allow for the lowest possible reservation of processor resources that meets all the constraints.

Another problem with this approach is that the static order scheduler does not necessarily produce the solution that allows for the smallest slice sizes. For cases, as in our example, where the execution time of the static order scheduler permits it, we can run the scheduler inside a loop that searches for lower slice sizes in a way similar to the binary and random slice allocators, with the difference that the whole static order schedule is recomputed instead of just the MCM, for each attempted combination of slice sizes.

5.6 Example

We will now show how to apply these scheduling techniques to an actual application in the Software-Defined Radio domain. Assume a multiprocessor system designed for baseband decoding. It includes a general-purpose core, an ARM, to handle control and generic functionality, a vector-processor core, the EVP [8], to handle detection, synchronization and demodulation, and an application-specific Software Codec processor that takes care of the baseband coding and decoding functions. All these processors are interconnected via an Æthe real Network-On-Chip [30]. The platform is used to handle several radio standards (Wireless LAN, TD-SCDMA, UMTS, DVP-H, DRM). In our example we will assume that we want to derive scheduler settings such that we can run Wireless LAN (WLAN) 802.11a and

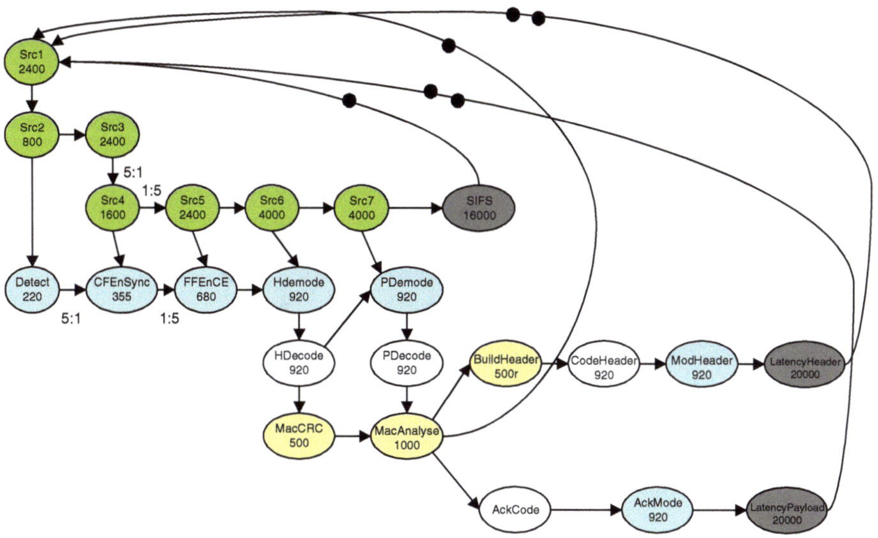

Fig. 5.12 A Wireless LAN 802.11a receiver job

TD-SCDMA simultaneously, with independent start and stop, and allow for up to two job instances to be active at a time, including configurations with two WLAN instances and two TD-SCDMA instances.

Figure 5.12 depicts the timed data-flow model of a WLAN 802.11a job. Execution times (indicated under the actor names) are given in nanoseconds. The different shading of nodes indicates the different cores to which the actors are assigned. Nodes with names starting by "Src" model the source (inputs from an external RF unit), the nodes "LatencyHeader" and "LatencyPayload" and their adjacent edges are used to convert latency into throughput constraints, as described in Chap. 4. The source is in this case sporadic, but we have seen in Chap. 4 that the maximum latency associated with a sporadic source can be bound by the maximum latency for a strictly periodic source with a rate equal to the minimal interval between activations of the sporadic source, and therefore we treat the source as periodic.

For space reasons, the Synchronization step is represented in Multi-Rate Data flow syntax: 5 "CFESync" nodes process the output of 5 "Src" nodes in a chain of sources and synchronization nodes. The number of "PayloadDemode" actors and respective sources may vary between 1 and 255. We only depict the case where there is only one "PayloadDemode" actor. Source and Latency actors are not scheduled. This graph has a required maximum production period of $\mu_D(\text{WLAN}) = 40{,}000$ ns.

Figure 5.13 depicts the timed data-flow graph model of a TD-SCDMA job. The "Rx" nodes represent the source. The "Latency" nodes and adjacent edges are used to convert latency into throughput requirements as described in Chap. 4 (we assume, in this case, a strictly periodic source). The required maximum production period is $\mu_D(\text{TDSCDMA}) = 675{,}000$ ns.

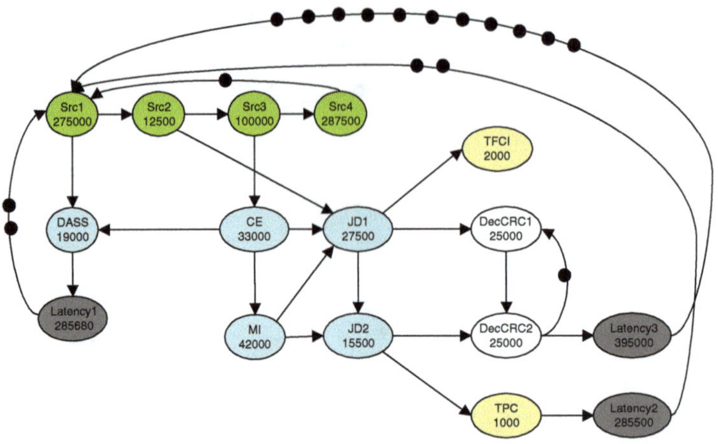

Fig. 5.13 A TD-SCDMA job

Table 5.2 WLAN scheduling results

Weights			Reservation (%)		
EVP	SwC	ARM	EVP	SwC	ARM
1	1	1	45	45	21
2	1	1	28	45	45
2	2	1	28	45	45
1	2	1	45	28	45

We start by choosing time wheel periods. We assume a worst-case context-switch time of 100 ns for all processors. We want to support two jobs at a time and we would like to spend less than 10 % of our time context-switching, so we set our wheel time to 2,000 ns. Therefore, $P(EVP) = 2{,}000$, $P(SWC) = 2{,}000$, $P(ARM) = 2{,}000$.

As we want to be able to run two WLAN or two TD-SCDMA jobs simultaneously, we set the maximum utilization per processor at 45 %, (half processor utilization minus context-switching overhead). That is, the maximum slice size for any processor is $\hat{S}(p) = 0.45 \times P(p)$.

For both jobs, we first computed static order schedules and then time slice extensions. In both cases, a static order schedule that meets the timing constraints (including the $\hat{S}(p)$ limitation on slice size per processor) was found. These schedules were used to calculate slice time optimizations. We varied the values of the costs of the three processors to search for trade-offs. It turns out that the algorithms tends heavily towards allocating all slack to a particular processor. A change in weights typically forces a drastic change from allocating slack to one processor to another, making this approach essentially equivalent to a binary search with weights associated with processor types.

In the case of the WLAN – the results are shown in Table 5.2 – varying the weights for each processor type allowed us to decrease the reservation of one of the processors from 45 to 28 %. The table pretty much exhausts the possible trade-offs.

Table 5.3 TD-SCDMA scheduling results

Weights			Reservation (%)		
EVP	SwC	ARM	EVP	SwC	ARM
1	1	1	45	15	2
100	1	1	43	15	2
1,000	1	1	33	15	45

Table 5.4 WLAN scheduling results with binary search slicer

Priority			Reservation (%)		
EVP	SwC	ARM	EVP	SwC	ARM
3	X	X	25	45	45
X	3	X	45	25	45
2	1	3	44	45	18
1	2	3	45	44	18

It suggests that it makes sense to keep a table with several configurations and allow the admission controller to chose different TDM settings for a job instance taking into account the current level of resource reservation.

The results for the TD-SCDMA are shown in Table 5.3. One thing to notice here is that the optimal slice times for the ARM and the Software Codec are obtained for equal weights. Increasing the weights of either ARM or Software Codec by any amount didn't further decrease their slice times. By comparing the results in the first and the second entries of this table, we observe one of the problems with our linear programming formulation: the optimized value of $N(ARM)$ in the first entry does not allow a smaller slice time than the much lower value obtained in the second entry; it does however prevent the slice time for the EVP to be decreased from 900 to 860. This example does not allow many trade-offs: to allow a decrease from 900 to 660 on the EVP, we have to increase its weight to 1,000. Moreover, this comes at the cost of changing the utilization of the ARM from 2 to 45 %. Although not perfect, our slice time optimization based on the linear programming formulation allows a decrease of the reservation to 15 % for the Software Codec and 2 % for the ARM, from the imposed maximum 45 %.

Since the execution times of the scheduler were very fast for these applications, ranging from a few milliseconds for the TD-SCDMA and for the WLAN when considering payloads of 1, to a little more than a minute for the largest WLAN payload size (256), we decided to run the scheduler within a loop, and search for improved slice sizes.

If we apply the binary search slicer algorithm on a loop where for each stage we run the static order scheduler, we obtain, for the WLAN transceiver, the results presented in Table 5.4, where 3 is the highest priority, 1 is the lowest, and X is any priority lower than 3.

For the TD-SCDMA example, we obtained the results presented in Table 5.5.

As for the randomized slicer, it yields the same result for TD-SCDMA as the binary slicer since, for this transceiver, the minimum slice values for each processor are seemingly independent of the others. This has to do with the fact that the cycle that becomes the most critical when we reduce the resource reservation for each

Table 5.5 TD-SCDMA scheduling results with randomized search slicer

Priority			Reservation (%)		
EVP	SwC	ARM	EVP	SwC	ARM
X	X	X	21	8	1

Table 5.6 WLAN scheduling results with binary search slicer

Priority			Reservation (%)		
EVP	SwC	ARM	EVP	SwC	ARM
1	1	1	33	36	36

of the processors is different. For the EVP, this is a cycle that includes the edge between Latency2 and Source1. For the Software Codec, it is the cycle between DecodeCRC1 and DecodeCRC2. As for the ARM, we obtained in both cases the lowest reservation we considered (1 %).

In the WLAN case, the randomized slicer yields other slice allocations depending on the weights given to the processor types. By giving the same weights to the three processor types, and using a step equal to 1 % of the period, we obtained a slice distribution such as the one shown in Table 5.6.

As for the buffer sizing, we ran the buffer sizing algorithm before and after static ordering and slicing, and obtained exactly the same results. This may be accounted for by the relatively few possibilities for pipelined execution within the graphs of these applications. Most radio applications we studied (UMTS, LTE, DVB-T, DVB-H), however, seem to have relatively similar structures.

The best choice for ordering the stages in the scheduler may be very dependent on the characteristics of the specific application and platform. In this case, the relatively small execution times of the static order scheduler allowed us to run a full search for schedules to optimize the slice sizes. The relatively small search space for optimal buffer allocation made moot the choice of when to perform buffer sizing. This may vary widely from application to application, requiring flexibility in the implementation of the scheduler. Since the temporal impact of all the decisions can be simply modeled in a data flow graph, all scheduling decisions can be imposed as pre-conditions to any stage of the scheduler, making it simple and easy to re-order the stages of the scheduler, or even merge them.

5.7 Further Reading

Much work has been published on the scheduling of data flow graphs with real-time requirements. The level of dynamicity we allow in the start and stop of jobs is what basically differentiates our problem from other multiprocessor real-time scheduling problems. We will review proposed approaches that could be adapted to solve our problem.

The most viable alternative to the solution proposed in this book is to use pre-compiled resource allocation configurations, such as in CPA [84]. For each job-mix, a separate optimal static schedule is derived at compile time and stored in a look-up table. During operation, when there is a request to start a job, the runtime system checks which jobs are active and selects the appropriate configuration. This approach is not without problems. First, a different configuration has to be stored for each combination of jobs, which means that the number of configurations grows exponentially with the number of jobs. Second, if a job is not known at design time, it will force a whole new set of configurations to be compiled later. Third, it is difficult to assure continuity of execution of already running jobs during reconfiguration. For continuity, a configuration should be generated for each transition from a job-mix to another which easily becomes infeasible because of the large amount of configurations that may need to be generated.

The technique proposed in this book calculates scheduling budgets per job at compile time. During runtime two distinct temporal phases alternate: configuration phase and steady-state execution phase. During configuration phases, resources are allocated to jobs; during steady-state resource allocation is fixed. This is comparable with semi-static techniques [16], where system execution is divided in phases and resource allocation is redone at the beginning of each phase. But while in semi-static systems phases are periodical, in the technique proposed in this chapter case reconfiguration phases are triggered by an external request to start or stop a job. Also, the technique presented in this chapter assumes that an actor, once placed, cannot be moved to another processor.

This strategy also has similarities with other time-multiplexing strategies such as gang scheduling [19]. Jobs correspond roughly to a 'gang', i.e. a group of tasks (actors in our case) with data dependencies. There is, however, an important difference: in gang scheduling, time-multiplexing is global, i.e., the temporal slots are uniform across all processors, and synchronized context-switching is required. In our case, time-multiplexing is local to each individual processing element. The strategy presented in this book has several advantages over gang scheduling: it does not require global synchronization of context switches, which hinders design scalability; it can leave less unused resources because its time sharing is more fine grained; and it also allows for a different scheduling mechanism per processor.

Stuijk et al. [86] describes a method with many similarities to the one we presented in this book, since it also mixes static order and TDM scheduling for homogeneous multiprocessor systems. It does not consider non-preemptive scheduling or heterogeneous systems, and it does not address different start/stop times. To account for the effect of TDM on the response times, it uses simulation of the self-timed execution of the data flow graph with worst-case response times, while keeping track of the state of the TDM wheel. This work also differs from this book in that actor to processor mapping and static ordering phases are decoupled. First an heuristic is applied to find a processor mapping, and then the static order schedule is determined by a ready-list scheduler with no backtracking and no timing constraints, which means that the static order schedule has one single chance at finding a feasible order. This has the advantage of considerably speeding up the execution time of the compile time scheduler. However, no explicit heuristic is proposed to decide which of two simultaneously ready actors mapped to the same processor is scheduled first. If the static order scheduler fails, the tool simply does a new mapping and tries again to statically order it. Slice minimization is done using binary search after static ordering, using data flow simulation to check for timing constraint violation at each step. In this approach, only TDM schedulers can be supported as timing analysis is dependent on the simulation of the TDM time-wheels.

5.8 Conclusion

In this chapter we present a scheduling strategy and a scheduling flow to solve the problem of running multiple jobs with different rates and different start/stop times on a heterogeneous multiprocessor. The scheduling strategy involves a combination of static order scheduling per job per processor, and TDM or Non-preemptive Non-blocking Round Robin scheduling to arbitrate between different jobs in each processor. We argued that the combination of dynamic scheduling with static order scheduling is desirable and can be modeled for the purpose of temporal analysis. We have shown how the temporal analysis model can serve as a basis for a scheduling flow. We have presented algorithms to estimate the temporal slack that can be allowed per actor for a timed SRDF, and how to exploit this time slack by decreasing slice times.

Although we present the flow using Time-Division Multiplex and Non-Preemptive Non-Blocking Round Robin for inter-job scheduling, this does not present an essential limitation of the techniques: any scheduler that can be represented by a latency-rate model can be employed, although changes may be required to the computation of scheduler settings. Also, any new data flow model of a scheduler can be easily supported by simply changing the model generation stage to create different actors and edges.

There are several open issues to address in future work. One is finding a heuristic approach to perform the clustering and static ordering of actors, since the current algorithm is untrackable and, although manageable in the relatively small graphs of the transceivers we studied, it does not scale easily to larger graphs (and larger graphs can arise from CSDF to SRDF conversion for even relatively small CSDF graphs). Furthermore, the current approach allows communication and buffer capacity to be taken into account as constraints, but it would also be of interest to search for scheduling settings that optimize for low communication requirements and/or small buffer sizes.

Chapter 6
Mode-Controlled Data Flow

In the process of choosing a Data Flow (DF) programming model for describing an application, one must make a difficult trade off between expressiveness and analytical properties. At one extreme, a Dynamic Data-flow (DDF) model is expressive enough to mimic the behavior of a Turing machine [10], but lacks many useful analytical properties; for instance, for an arbitrary DDF graph it may be impossible to verify if it is free of deadlocks, or if it can execute for indefinite time on bounded buffer space [10]. On the other hand, Static Data flow (StDF) variants (such as Multi-Rate DF [57], Single-Rate DF [80], or Cyclo-Static DF [9]) allow for powerful analysis, such as the verification of deadlock-freedom, determination of maximum achievable throughput, and verification of latency and throughput constraints, but have limited expressivity: all actors must work with fixed data rates, i.e., all quantities of data sent/received per actor firing cannot be dependent on the values of the input data. Because of these limitations, StDF models tend to be reserved for application domains where task activation is data-driven, data rates regular, and real-time guarantees required.

For the most, this is the case for Base-Band Software-Defined Radio (SDR) [8]. On a typical digital baseband algorithm, data flows through three processing stages: filtering, modulation/demodulation and coding/encoding. Broadcast standards typically have throughput requirements that must be met, while bi-directional communication standards also impose maximum latencies.

StDF computation models have been proposed [62] for usage in this domain. A problem with such approach is that some SDR applications are subject to occasional changes of operation mode, i.e. changes in the sequence of algorithms that must be executed to process the incoming stream. A DVB-H receiver, for example, executes the same demodulation-decoding flow of streaming tasks during most of its operation, but during initialization and occasionally during its active state it needs to execute a different sequence of tasks in order to achieve synchronization with the incoming stream. This synchronization mode takes an unknown number of iterations to succeed. Therefore, while static data rates can be used to represent the

O. Moreira and H. Corporaal, *Scheduling Real-Time Streaming Applications onto an Embedded Multiprocessor*, Embedded Systems 24, DOI 10.1007/978-3-319-01246-9_6,

message passing within each synchronization attempt, the same cannot be done to represent the dependencies and data communication between synchronization and decoding modes.

A more severe problem can be observed in Wireless LAN (WLAN). On a WLAN 11a receiver, packet detection is performed as a loop in the RF interface until it succeeds. Then, the baseband starts executing. First, synchronization is performed. If this is successful, a demode-decode sequence is executed for each received symbol, until the variable-sized payload of the packet is fully processed. Each of these operation modes can be expressed as an (analyzable) StDF graph. The change from mode to mode, however, cannot be expressed within the confines of StDF. In terms of DF, a change of operation mode is more like a change of the computation graph. However, for SDR applications, hard real-time requirements may be defined across operation mode transitions and therefore temporal analysis is only useful if mode transitions can be taken into account. Such is the case with the WLAN 11a example, where there is a maximum allowed latency from the time when the packet starts to be sent and the moment when the receiver sends an acknowledgment of packet reception to the sender. To make the problem more complex, there are data dependencies between modes, such that the execution of a mode is conditioned by the schedule of the previous one.

The behavior during one of these mode transitions cannot be conveniently expressed in StDF. A solution that we have used in the past is for an analysis specialist to design an StDF model that represents worst-case assumptions about the execution of a given, concurrent WLAN receiver implementation that does not necessarily conform to the strict StDF construction rules. The StDF model can then be used for temporal analysis and scheduling purposes. One such model was used for our WLAN 11a example in Chap. 5.

This approach is slow, difficult, and requires both data-flow modeling expertise and in-depth knowledge of the application. Furthermore, it is impossible to guarantee that the model does indeed conform to the actual implementation. What we would like to have is a programming model that enables the automatic extraction of an analysis model. However, the programming model has to be powerful enough to describe these mode transitions and still allow for hard-real-time analysis.

In this chapter we propose a DF programming model that can be seen as a restriction of Boolean DF. It allows for mode switching without compromising the properties that make the graph amenable to temporal analysis, while keeping the programming model as simple as possible. For need of a name, we refer to this restricted model as Mode-Controlled DF (MCDF). The chapter is organized as follows: in the next section, we present an overview of MCDF; in the following section, we describe in detail the new types of data flow actors introduced by MCDF; in Sect. 6.3 we present the rules to construct an analyzable (i.e., well-constructed) MCDF graph; Sect. 6.4 shows how radios can be modeled in MCDF, by providing two examples, a DVB-T receiver and a WLAN 802.11a receiver; Sect. 6.5 discusses the basic analytical properties of well-constructed MCDF graphs; Sect. 6.6 provides techniques for temporal analysis of MCDF graphs; Sect. 6.7 describes the scheduling techniques for MCDF graphs; Sect. 6.8 describes the results of scheduling a WLAN receiver modeled as an MCDF graph on a multiprocessor platform; Sect. 6.9 suggests further reading on this topic; and Sect. 6.10 concludes the chapter.

6.1 Model Overview

Intuitively, a Mode-Controlled DF graph is a data flow graph where for each firing of a designated actor, called the Mode Controller (MC), only actors belonging to a specific, pre-defined subgraph of the complete MCDF graph are fired. A specific subgraph is chosen for firing, depending on an output value produced by the Mode Controller. Therefore there are data-dependent actor firings. After all actors in the chosen subgraph have fired, the graph returns to the initial token distribution. An iteration of an MCDF graph corresponds to the firing of all actors in one of these subgraphs. The model allows verification that no deadlocks can occur and behavior is kept deterministic in the Kahn Process Network [48] sense, i.e., provided that FIFO communication is blocking, and that the internal functionality of actors has no notion of time, the outcome of the computations is independent of the time at which the actors are activated.

6.2 MCDF Constructs

In this section we will describe the main building blocks of a Mode-Controlled DF graph.

6.2.1 *The Mode Controller*

The Mode Controller (MC) is a special node in MCDF. The special characteristic of the MC is that one of its outputs drives the control input ports of all data-dependent actors in the MCDF graph. We refer to this output port as the **mode control** port. At each firing, the MC produces a single token in the mode control port, and there are no delays on the edges that connect the mode control port to the control input ports of all data-dependent actors.

For any given MCDF graph, there is a fixed number of modes, M. The value of the token produced by the MC in the Mode Control port has to be an integer within the closed interval from 1 to M. This value represents the mode selected for the current iteration of the MCDF graph. Tokens produced through the mode control port are referred to as **control tokens**.

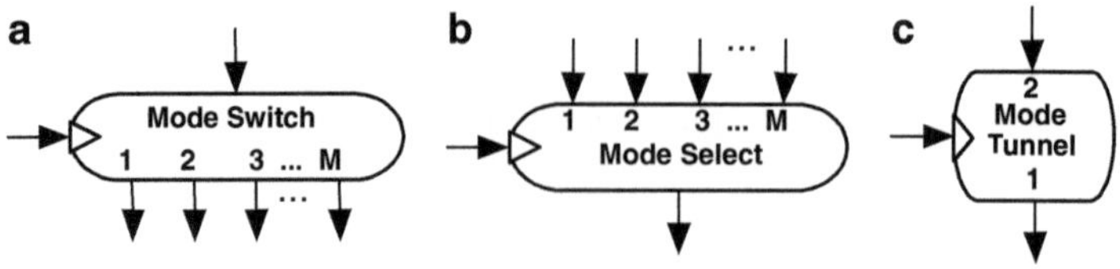

Fig. 6.1 Variable rate nodes in MCDF

6.2.2 Data-Dependent Actors

Besides SRDF actors, an MCDF graph allows the usage of three types of data-dependent actors. These are the Mode Switch, Mode Select and Mode Tunnel actors. However, as we shall see, the Mode Tunnel can be represented by using both a Mode Switch and a Mode Select actors, and should be regarded as a convenient short-hand notation that we offer to the programmer to allow simpler specification of inter-mode communication.

The data-dependent actors have in common the fact that they all have a control input port. A control token is read from this port for each firing of the data-dependent actor. That value determines which other ports of the actor are producing/consuming data during this firing. Control tokens are produced and consumed as normal tokens.

Definition 6.1. A port is said to be **associated** with mode m if the consumption of a token with value m on the control input port causes data to be produced/consumed on this port, while the consumption of any other value will not cause data to be produced/consumed on this port.

Definition 6.2. A **Mode Switch** actor (Fig. 6.1a) has, besides the control input port, one data input port and M output ports. Each output port is associated with a mode. When a token is present on the control input port and on the data input port, the Mode Switch actor is fired. It consumes both input tokens and produces a token in the output port associated with the Mode indicated by the token consumed on the control input port. The output token has the same size and value as the token consumed on the data input port.

Definition 6.3. A **Mode Select** actor (Fig. 6.1b) has, besides the control input port, M data input ports and one output port. Each input port is associated with a mode. When a token is present on the control input port, its value is read and used to decide from which input port to consume a token. The actor is fired when a token is present in the control input port and in the data input port associated with the mode indicated by the token in the control input port. When fired, it consumes both of these tokens. At the end of the firing, it produces on the output port a token with the same size and value as read from the modal input port. If the input port was not connected, the output token will have some pre-defined default value, which can be mode dependent. Note that, as in the BDF Select, the firing rule must be evaluated in two steps.

Besides these two canonical types of data-dependent actors, we introduce a third type, the Mode Tunnel. Mode tunnels are in fact a shorthand for a construct involving both a Mode Switch and a Mode Select actor, and for that reason, they will be absent on our study of the properties of MCDF graphs. They are, however a very useful construct for programmers, as they allow the communication between different modes.

Definition 6.4. A **Mode Tunnel** actor (Fig. 6.1c) has, besides the control input port, one data input port and one data output port. The data input port is associated with an arbitrary mode m of the graph, and the data output port is associated with a mode n of the graph, different from m. When the token read at the control input port has the value m, the Mode Tunnel fires and consumes a token from the control input port and from the data input port. It stores the token read from the data input port in its internal state. When the control input port has value n, the Mode Tunnel fires, consuming that value and copying the token stored in its internal state to the data output port. The initial value of the internal state is graph-specific. In this way, the Mode Tunnel always delivers to its consumer the value produced by the last previous execution of the source orchestrated by the Mode Controller. As with Select and Mode Select, the firing rule of Mode Tunnel must be evaluated in two steps. We refer to a Mode Tunnel as an **m-to-n Tunnel**, if its input port is associated with mode m and its output port is associated with mode n.

Any of the data-dependent actors can be used without connecting all of its ports. In that case, an unconnected input port is considered to automatically meet its firing rules and an unconnected output port produces tokens that are lost. If the input port of a Mode Switch or the input port of a Mode Select associated with the mode selected for the presented iteration is unconnected, then the output for that firing of the data-dependent actor is a token with a pre-defined value.

6.2.3 Mode Tunnel Conversion

A Mode Tunnel can be represented as an MCDF construct containing one Mode Switch, one Mode Select and two other actors, as represented in Fig. 6.2 for a mode 1 to mode 2 tunnel. The edge labeled as "data in" represents the data input arc of the Mode Tunnel. The edge labeled as "data out" represents the data output arc of the Mode Tunnel. The two "control in" arcs receive the value in the control input arc of the Mode Tunnel. When fired, actor "a", connected to the data input arc, copies the value received from the data input arc to its output arc. Actor "b" copies the value received from Switch to its output arcs. The arc that connects the output of the Mode Select to the input of the Mode Switch has a delay of one, and it is here that the last datum read from the input arc of the Mode Tunnel is stored. The example depicts the case where there are only two modes. For other modes not involved in the tunnel, a direct edge with zero delay would connect the output of the Mode Switch to the input of the Mode Select. From here on, we will not refer to Mode Tunnels when

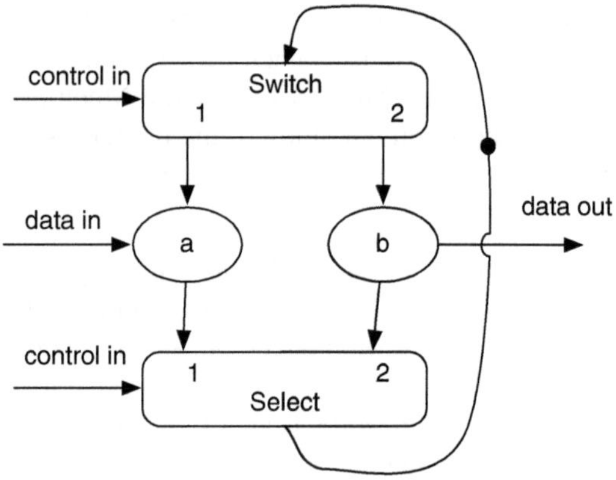

Fig. 6.2 Construct to represent Mode Tunnel in terms of Mode Switch, Mode Select and SRDF actors

discussing MCDF properties, analysis and scheduling, as we will assume that these have been previously converted into this construct, but we will use it in describing applications modeled as MCDF, as it greatly simplifies models where values must be communicated between modal actors from different modes.

6.3 MCDF Construction Rules

The rules we will present for the construction of a well-constructed MCDF graph are designed to guarantee that it can always return to the initial state of token placements independently of the sequence of mode control tokens generated by the Mode Controller. More specifically, we want to ensure that each firing of the MC enables all actors pertaining to a mode and all actors not specifically pertaining to any mode to fire until all edges return to the initial state, without any other activation of the Mode Controller being required. Here we will assume that all actors that do not have data-dependent behavior are single-rate. In Sect. 6.9 we will discuss how the model can be extended to multi-rate actors.

An MCDF graph with M modes is composed of:

- A Mode Control actor (MC);
- An arbitrary number of Mode Select, Mode Switch actors;
- An arbitrary number of static, single-rate actors

Now, we will define some terminology which will help us specify the construction rules.

Definition 6.5. Two ports are said to be **connected** if and only if there is an arc between them in the graph. In the same way, two actors are said to be connected if and only if there is an input port of one of them and an output port of the other that are connected.

Definition 6.6. An edge is said to be **delay-less** if its delay valuation is zero. A path is said to be delay-less if all the edges that it traverses are delay-less.

Definition 6.7 (Actor Modality). Actors in an MCDF graph are annotated with a valuation *mode:* $V \rightarrow \{1, \ldots M\}$. Since only some actors execute for specific modes, *mode* is a partial function. If *mode(i)* is defined, then the actor is said to be **modal** and we can say that actor i **belongs** to *mode(i)*. If *mode(i)* is undefined, denoted by *mode(i)*$= \bot$, the actor is said to be **amodal**.

Please note that modal actors do not have a mode control port. The only actors to have a mode control port are the data dependent actors, Mode Switch and Mode Select (and Mode Tunnel), which are always amodal, i.e. are fired in all iterations, independent of the value of the mode control token. The Mode Controller itself is amodal, and its firing condition is data-value independent.

Definition 6.8 (Actor Type). There are some actors in MCDF that have special attributes, these are the Mode Controller, the Mode Switch and the Mode Select. We use valuation *atype:* $V \rightarrow \{mc, switch, select, normal\}$, to indicate any special attributes of an actor.

An MCDF graph is defined by the tuple $G = (V, E, t, d, M, mode, atype)$, where V is the set of actors, E is the set of edges, t is the timing valuation of actors, d is the integer-value delay of edges, and M is the number of modes in the graph, *mode* is the mode valuation of actors and *atype* is the actor type valuation for actors.

Definition 6.9. The **Mode** m **Subgraph of** G is a sub-graph $G_m = (V_m, E_m, t, d)$ of MCDF graph $G = (V, E, t, d, M, mode, atype)$, such that its vertex set V_m is composed of all amodal actors, and all actors that belong to mode m, and its edge set is composed of all edges which are in E and whose sources and sinks both belong to V_m, and where t and d are restricted to V_m.

To simplify the notation, we will not explicitly model ports. We follow a number of assumptions about ports: if a modal node is connected to a data-dependent actor such as a Mode Switch or Mode Select, it is always to a port associated with its mode. Furthermore, an output port can be connected to multiple edges – i.e. the value produced through that port is produced onto all edges connected to it – such a port can be referred to as a multi-port. The mode control port is a multi-port if there are more than one data-dependent actors (i.e. Mode Switches and Mode Selects). An input port can only be connected to a single edge.

The construction rules that must be respected by a MCDF graph for it to be considered well-constructed are as follows:

Rule 1. *There is only one mode controller.*

Rule 2. *Modal actors can either be connected to other modal actors, as sinks to the output ports associated with their mode on Mode Switches, or as sources to the input ports associated with their mode on Mode Selects. On the other hand, the ports of amodal actors other than the output ports of Mode Switches and the input ports of Mode Selects can only be connected to fixed rate ports of other amodal actors. This means that if G is an MCDF graph it must hold for edge* $(i, j) \in E$ *that*

$$(mode(i) = mode(j)) \vee (atype(i) = switch \wedge mode(j) \neq \bot)$$
$$\vee (atype(j) = select \wedge mode(i) \neq \bot)$$

Rule 3. *The mode control output port of the Mode Controller is connected to all control input ports in the graph through delay-less edges. This means that for any edge $(i, j) \in V$ where $atype(i) = mc$ it must hold that*

$$d(i, j) = 0 \wedge (atype(j) = switch \vee atype(j) = select) \tag{6.1}$$

Rule 4. *There are no delay-less cycles in the graph.*

Rule 3 ensures there is a single condition that drives all data-dependent actors. Furthermore, it guarantees that for the same firing count the same mode control token is read by all data-dependent actors, i.e. if the nth firing of a given data-dependent actor activates its ports associated with mode m, then we can expect the nth firing of another data-dependent actor in the graph to activate its own ports associated with mode m.

The concept of iteration for MCDF is the same as for SRDF: a sequence of actor firings that brings the graph back to the initial token distribution. The difference is that in MCDF not all actors fire in each iteration. In fact, as we shall prove, only amodal actors fire once per each iteration, and the count of completed iterations of the graph is equal to the number of firings of MC.

Rule 4 is needed to obtain deadlock-free graphs. A cycle with no delays would create a cyclic dependency between firings on the same iteration which could eventually deadlock the execution of the graph.

6.4 Radio Modeling in MCDF

MCDF was designed to allow the easy expression of modal behavior in transceivers. In this section, we will provide two examples of how radios can be modeled in MCDF: a DVB-T receiver and a Wireless LAN 11a receiver.

6.4.1 Example Application: DVB-T Receiver

Figure 6.3 depicts our MCDF model for a DVB-T receiver application. The DVB-T receiver requires an MCDF graph with three modes: mode 1 handles synchronization, mode 2 simply drops input data and mode 3 demodulates and decodes the input data and sends it to the output stream. Both in synchronization and demodulation modes, the DVB-T receiver must regulate the timing and frequency of the receiver analog front-end. This is represented by an analog control that is issued by the "sync" actor in mode 1 and by the "dem" actor in mode 3, that is sent through a mode select back to the source. As for the operation of the mode

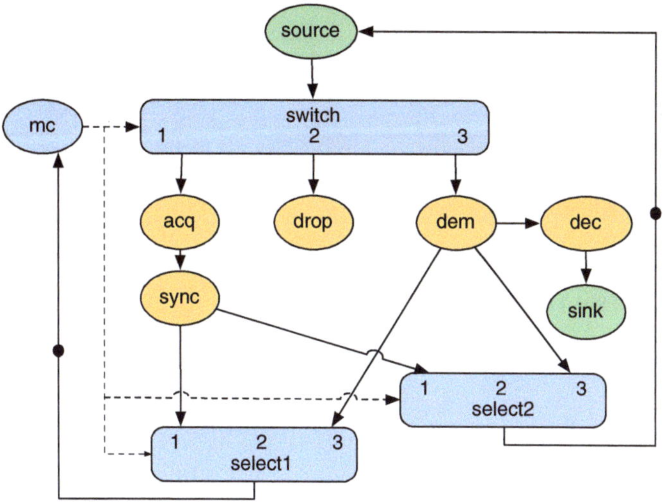

Fig. 6.3 MCDF graph for a DVB-T receiver

controller, it selects, in the first iteration, the synchronization mode. It will continue to operate in this mode until synchronization with the incoming stream is detected. In that case, the receiver must align itself with the beginning of a DVB-T frame, and, for a number of rounds determined by the outcome of synchronization, it will drop input data. It will then go into demodulation mode (mode 3), where it will stay until either the receiver is stopped or synchronization is lost, in which case it will go back to the synchronization mode. The requirement for this receiver is that no matter what mode it is running, it keeps up with the period of the source (about 900 μs).

6.4.2 Example Application: Wireless LAN Receiver

Figure 6.4 depicts the timing of a WLAN 802.11a packet. In a WLAN 802.11a receiver, baseband processing starts once a packet is detected at the RF. First, synchronization is performed over the first 8 μs. These are two groups of five short symbols – five short symbols are equivalent to one long symbol. If this is successful, the packet header is demodulated and decoded to determine the size of the payload. After this, the OFDM symbols of the variable-sized payload – from 1 to 256 OFDM symbols, each with a length of 4 μs – are processed one by one by a demode-decode loop, followed by a Cyclic-Redundancy Check (CRC). If CRC is successful, an acknowledgment packet must be sent within 16 μs of the end of reception. This time guard of 16 μs is known as the Short Intra-Frame Spacing (SIFS).

Figure 6.5 depicts our MCDF model for a Wireless LAN receiver application. Arcs that communicate control tokens are depicted by dashed lines. The graph has four modes: Synchronization (mode 1), Header Processing (mode 2), Payload

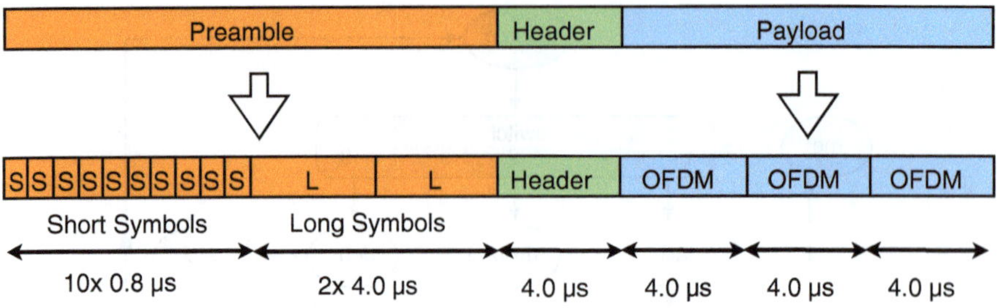

Fig. 6.4 WLAN 802.11a packet structure

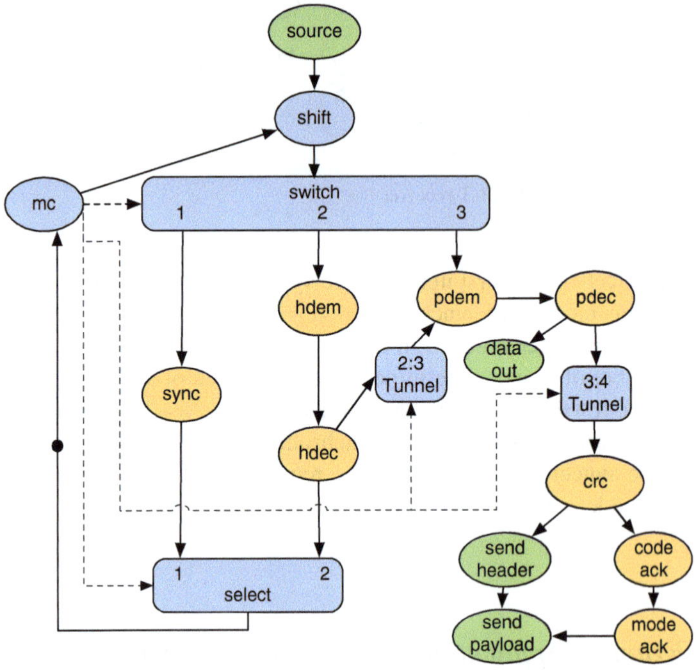

Fig. 6.5 MCDF graph for a WLAN 11a receiver

Processing (mode 3) and CRC (mode 4). The input from the RF source ("source" actor) is sent to the "shift" actor, which performs alignment at the start of each OFDM symbol, by pre-pending the end of the previous token to the current one, and storing the remainder of the current token to pre-pend to the next. The length of the prefix is first set to 0, and is later determined during synchronization, and transmitted to "shift" through the Mode Controller, that received it from the Mode Select "select" actor. From "shift", data is sent through the Mode Switch "switch" to the mode selected by "mc." The decoding of a packet starts with "mc" selecting mode 1. At the end of each firing, "sync" informs "mc" through "select" whether it succeeded to synchronize. Depending on the outcome, "mc" will choose whether to fire mode 1 again, or start processing the header. Also,"mc" sends to "shift" the current value of the offset, as given by the return value of mode 1, received from "select". Mode 2 fires two modal actors: one on the EVP ("hdem") and the other on the software codec ("hdec"). The status sent by "hdec" to "mc" via "select" indicates the length of the payload. Also, through a Tunnel from mode 2 to 3, several demodulation parameters for this message are communicated to the "pdem" actor that does the demodulation of payload symbols in mode 3. For a number of firings equal to the length of the payload, "mc" will select mode 3, firing actor pdem in the EVP followed by actor "pdec" in the Software Codec. Once the full payload has been received, "mc" will select mode 4 for execution, activating the consumer side of the Tunnel between actors "pdec" and "crc". The "crc" actor executes in the ARM processor. After the CRC check is completed, a response message is encoded in the Software Codec (actor "code ack"), modulated in the EVP (actor "mode ack"), and sent to the base-station through the RF. Notice that only when modes 1 or 2 are selected is the decision of the Mode Controller dependent on the last input data. For modes 3 and 4, decisions can be taken ahead of this, and in the graph this is reflected by the fact that "select" can immediately generate a data token to fire again "mc" without waiting for any of the modal actors. Notice also that for the execution where "mc" selects mode 4, data from the source is discarded, since the output port of "switch" for mode 4 is not connected.

In the next iteration, "mc" reverts to its initial state and the graph is ready to process a new incoming packet.

6.5 Properties

In this section we will discuss relevant analytical properties of MCDF graphs.

Some important temporal properties of Mode-Controlled Data-Flow stem from two facts: (1) MCDF is a subset of Integer-Controlled DF (IDF) ([10] – see Chap. 3), and therefore inherits all its properties; (2) a tight worst-case timing model of its execution can be built that conforms to the SRDF model and allows us to transpose some SRDF properties to MCDF graphs.

Other properties stem from the choice we made on defining the construction rules. In the following sections, we will present some important properties of MCDF, and provide proofs that they hold.

6.5.1 Notation

Consider an MCDF graph $G = (V, E, t, d, M, mode, atype)$, where V is the set of actors, E is the set of edges, t is the timing valuation of actors, d is the integer-value delay of an edge, and M is the number of modes in the graph, *mode* is the mode valuation of actors and *atype* is the actor type valuation for actors.

Let a sequence $c = [m_1, m_2, \ldots, m_n]$, with $m_k \in \{1, \ldots, M\}$ represent the mode control sequence, i.e., the sequence of values of the tokens produced by the Mode Controller on a given execution of the MCDF graph. Let $c(k)$ return the value of the tokens produced by the mc node on its $(k+1)$th firing, or the kth element of c.

6.5.2 *Determinism*

Any MCDF graph is also an IDF graph, and has all the properties of IDF graphs. Therefore, it is **deterministic**, in the same sense as Kahn's Process Networks [48]: given the same input history (i.e. given a specific sequence of values produced by the firings of the source actor), an MCDF graph always produces the same exact output, independently of the time at which each actor firing happens (note that this is valid when we assume FIFO ordering, as described in Chap. 3).

6.5.3 *Linear Timing*

Another interesting property of MCDF is the **linear timing** of its schedules. We proved that this property holds for self-timed schedules of SRDF graphs in Chap. 4. Take a self-timed schedule $s(i,k)$ of a data-flow graph $G = (V, N, t, d)$. Then, if the $(n+1)$th firing of a given actor $a \in N$ is delayed by a time interval of Δ such that its new firing time is $s'(a,n) = s(a,n) + \Delta$, then s' is guaranteed to be an admissible schedule of the graph if, for all firings dependent on firing (a,n), $s'(i,j) = s(i,j) + \Delta$, while for firings independent of (a,n), $s(i,k) = s'(i,k)$. Note that in SRDF an actor firing (i,k) is dependent on (a,n) if, by recursively applying Eq. 4.2 we find an expression dependent on $s(a,n)$. This property holds because the firing rule for the activation of an SRDF actor is lower-bounded by a max function: when all the inputs are available, the actor can fire. If an input is available later than predicted, then the actor can, in the worst-case, be delayed by the same amount of time. In the best case, the input still arrives in time for the firing of the actor to be possible at the predicted time. Conversely, if a start time is anticipated by a time interval of Δ (assuming that no dependencies are broken, for instance, because the execution time of a producer was shortened), one can only expect any other firing dependent on that firing to happen at most Δ earlier than predicted, and never later. Since MCDF is deterministic, the sequence of mode control signals does not change when a given firing is delayed, and, for a given fixed sequence of mode control signals, since the firing rule of an MCDF actor is also based on a max rule on the arrival of inputs, the theorem will still hold for an MCDF graph.

6.5.4 *Iterative Behavior and Deadlock Freedom*

In this section, we show that a well-constructed MCDF graph is deadlock-free, and that its execution exhibits a behavior where the graph always returns to the initial token distribution after a well-defined firing sequence, by stating and proving the following theorem:

Theorem 6.1. *For a well-constructed Mode-Controlled Data Flow graph* $G = (V, E, t, d, M, mode, atype)$ *there is a non-empty firing sequence for each possible value of the mode control token that will bring the graph back to its original state (i.e. the initial token distribution). In each such firing sequence, all the amodal nodes fire once, all the modal nodes of a mode m selected by the Mode Controller fire once, and all other modal nodes do not fire.*

Proof. We will first compute how many firings of each actor are necessary to bring the graph back to its initial state. Let $r(i)$ denote the number of times that actor i must be fired on a firing sequence to bring the graph back to the initial token distribution. Then, for all edges $(i, j) \in E$ it must hold that at the end of the firing sequence the net increase of tokens in (i, j) is zero, i.e. per FIFO, for each token produced, a token has been consumed. This is normally expressed by the balance equations [57]:

$$\forall (i, j) \in E, r(i) \cdot prod(i, j) = r(j) \cdot cons(i, j) \tag{6.2}$$

where $prod(i, j)$ and $cons(i, j)$ represent, respectively, the number of tokens produced/consumed on the edge per firing of the producer/consumer.

On an MCDF graph, by construction, any non-data-dependent actor produces/-consumes exactly one token per output/input port.

For data-dependent actors (switches and selects) we need to represent symbolically the productions and consumptions on data-dependent ports. We use a similar convention as Buck [10] for representing IDF consumption and production rates for variable-rate ports. Then $p(c, j)$ designates the number of control tokens on a given mode control sequence c whose value is j, with $j \in \{1, 2, \ldots, M\}$.

We can categorize the edges of an MCDF graph, and try to solve the balance equations with respect to the number of repetitions $r(i)$ symbolically. There are four types of edges to be considered:

1. **Edges** (i, j) **between amodal nodes**: The balance equation is $r(i) = r(j)$, since $prod(i) = cons(j) = 1$ by construction.
2. **Edges** (i, j) **between modal nodes**: As in the previous case, the balance equation becomes $r(i) = r(j)$.
3. **Edges** (w, j) **from a switch actor** w **to a modal actor** j**:** Since w will produce a token in the output port associated with mode m every time it reads a mode control token with value m, if *mode*(j) indicates the mode value associated with the variable-rate port to which j is connected, then the balance equation becomes $p(c, mode(j)).r(w) = r(j)$.

4. **Edges (i, s) from a modal actor i to a select actor s:** Since s will consume a token in the input port associated with mode m every time it reads a mode control token with value m, if $mode(i)$ indicates the mode value associated with the variable-rate port to which i is connected, then the balance equation becomes $r(i) = p(c, mode(i)).r(j)$.

This is an exhaustive enumeration since, by construction, switches and selects must be amodal, and the construction rules do not allow edges between modal actors and constant-rate amodal actors.

Lets now assume a valid mode sequence $c = [m]$. According to our hypothesis, a solution of the balance equations must exist such that:

$$\begin{aligned} r(i) &= 1, && \text{if } mode(i) = \bot \\ r(i) &= 1, && \text{if } mode(i) = m \\ r(i) &= 0, && \text{if } mode(i) \neq m \wedge mode(i) \neq \bot \end{aligned}$$

It is easy to verify that this solution satisfies the balance equations for type 1 edges. It also satisfies the equations for type 2 edges, since these are, by construction, either between edges of mode m, in which case $r(i) = r(j) = 1$, or between edges of a mode $p \neq m$, and, in that case, $r(i) = r(j) = 0$. For type 3 edges, we have two cases: (1) if $mode(j) = m$, then $p(c, mode(j)) = 1$, and the balance equation becomes $r(w) = r(j) = 1$, which is satisfied by our solution, since j is an actor of mode m, or (2) $mode(j) \neq m$, and $p(c, mode(j)) = 0$, and the balance equation becomes $0.r(w) = r(j) = 0$, which again is satisfied by our solution. For type 4 edges, the argument is similar as for type 3 edges: there are two cases: (1) $mode(i) = m$, then $p(c, mode(i)) = 1$ and the balance equation becomes $r(i) = r(s) = 1$, where both values of $r(i)$ and $r(s)$ are satisfied by our solution and (2) $mode(i) \neq m$, then $p(c, mode(i)) = 0$ and the balance equation becomes $r(i) = 0.r(s) = 0$, which is satisfied by our solution. Therefore, our solution satisfies all balance equations on a well-constructed MCDF. □

Please notice that a consequence of this theorem is that to each firing of the MC and to each mode control value produced corresponds one iteration of the graph.

Theorem 6.2. *A well-constructed MCDF graph is deadlock-free.*

Proof. We only need to show that the graph will not deadlock during one iteration since, by definition, the graph will return to the initial state after concluding a full iteration.

We have already established that every switch actor fires once per iteration. For such one firing, it acts as a single-rate actor that produces one token at the output port associated with the mode m selected for the current firing. In the same way, a select actor acts like a single-rate actor that consumes one token at its input port associated with mode m.

Furthermore, starting from the initial token distribution, the MC can fire before any data-dependent actor, since there is a delay-less path from mc to all switches and selects, and the graph has no delay-less cycles. Therefore, for the current iteration, the firing of mc is only dependent on initial tokens and the firing of amodal nodes that cannot be reached through a delay-less path from modal nodes, again, because the graph has no delay-less cycles. For this reason, it is guaranteed that mc is able to fire at least once.

An SRDF graph G_m can be built that represents the execution of the current iteration of the original MCDF graph G. The vertex set N_m of this graph is composed of all amodal actors and actors belonging to mode m. Its edge set E_m is composed of all edges in the original MCDF graph whose sources and sinks belong to N_m.

In [81], it is proven that an SRDF graph is deadlock-free if it does not have any delay-less cycle. According to construction Rule 4, G_m has no delay-less cycles. Since G_m is a subgraph of G, it cannot have any delay-less cycles and is thus deadlock-free. Therefore, G does not deadlock when executing the iteration corresponding to the mode sequence $\{m\}$, for any $m \in \{1, \ldots, M\}$. Any iteration brings the graph back to the initial state. Since G does not deadlock for any single iteration, it cannot deadlock for any mode sequence, even if the execution of iterations is pipelined. □

6.5.5 FIFO Ordering, Firing and Iteration Counts

In SRDF graphs where all actors respect FIFO ordering of firings, the firing count of a particular firing of an actor indicates to which iteration it belongs. That is, the first firing of an actor will belong to the first iteration of the graph, the second firing of the actor to the second iteration of the graph, and so on. This is because, due to FIFO ordering, the tokens produced by an actor follow the same order as the tokens consumed by that actor, and every actor fires exactly once for each iteration, consuming exactly a token from each of its input FIFOs, and producing exactly a token on each of its output FIFOs.

As for SRDF graphs, we will only accept as well-constructed, MCDF graphs that respect FIFO ordering of actors. In MCDF, however, the count of iterations of the graph does not match the count of the number of firings for all modal actors, since they only fire in iterations where their mode has been selected by the MC. Nonetheless, for each modal actor, a firing belonging to iteration n can only happen after a firing of the same actor belonging to an iteration k, if $k < n$. This is because all actors respect FIFO ordering.

6.6 Schedules and Temporal Analysis

We will represent MCDF graph schedules as we did for SRDF graphs by employing a function that gives the start time of an actor for a specific iteration of the graph. There are however, two major differences. First, the schedule is now dependent on the mode sequence c. Second, the schedule is now a partial function, since, as we have seen in the previous section, for each graph iteration, only amodal actors and actors belonging to the particular mode that was selected for that iteration fired.

We will therefore represent an MCDF schedule as a partial function $s(i, k, c)$, where $i \in V$ is an actor, $k \in \mathbb{N}_0$ is the number of this iteration of the MCDF graph, starting from 0 for the first iteration, and c is the mode sequence. This is a partial function since the start time $s(i, k, c)$ is only defined if $mode(i) = c(k)$ or $mode(i) = \perp$, where $c(k)$ represents the $(k + 1)$th element of mode sequence c. If $mode(i) \neq c(k) \wedge mode(i) \neq \perp$, then $s(i, k, c)$ is undefined, or $s(i, k, c) = \perp$.

In order to present the precedence constraints for MCDF graphs, we need to determine on what firing of a producing modal actor is the firing of its consuming modal actor dependent. If the delay of the edge between producer and consumer is 0 then it is the same firing. If the delay of the edge is greater than 0 then it is dependent on the mode sequence. If, for instance, the delay is one, then the value produced by the first firing of the producer is consummed by the second firing of the consumer, and so on. However, the first firing of the consumer does not necessarily happen in the iteration before the second firing of the producer, since the mode of the two actors is not necessarily selected for two consecutive iterations. Given the iteration at which the second firing of the consummer occurred, we must find in the mode sequence the previous iteration at which the same mode was selected to identify the related firing of the producer. In general, given the initial delay of the edge and the mode sequence, we must be able to count back, for a specific iteration, an amount of firings of the producer equal to the delay of the edge to determine what was the iteration at which the value consummed in the present iteration was produced. We represent this iteration dependence by the following definition:

Definition 6.10 (Modal Delay δ). Let $G = (V, E, d, t, M, amode, atype)$ be a well-constructed MCDF graph. Let c be a valid mode sequence for graph G. Let n be the number of times mode $c(k)$ has been selected from iteration 0 to iteration k of an execution of G, for a mode sequence c. Let $k' \geq 0$ be an integer such that $c(k) = c(k')$ and the number of times mode $c(k)$ has been selected between iterations 0 and k' is $n - y$, where y is a positive integer. If k' exists, then the **modal delay** $\delta(k, c, y) = k - k'$. If k' does not exist, then by definition, we will say that $\delta(k, c, y) = k + y + 1$.

Lemma 6.1. *For any given integer $k \geq 0$, mode sequence c, and integer $y \geq 0$, it holds that $\delta(k, c, y) \geq y$.*

Proof. If k' as defined in Definition 6.10 does not exist, then $\delta(k, c, y) = k + y + 1 > d$, since $k \geq 0$. Assume that $c(k'') = m$ for any $k'' \in \mathbb{N}_0$ and $k'' \in [k - y, k]$. Under this assumption $k' = k - y' \Leftrightarrow k - k' = y \Leftrightarrow \delta(k, c, y) = d$. If mode m is not selected for all iterations in the mode sequence from $c(k - y)$ to $c(k)$, then it must be that $k' < k - y \Leftrightarrow k - k' > y \Leftrightarrow \delta(k, c, y) > y$. □

Another useful definition is the modality of an edge. In MCDF, the state of any edge $(i, j) \in E$ where either $mode(i) \neq \perp$ or $mode(j) \neq \perp$ only changes during the execution of an iteration k where $c(k) = mode(i)$ or $c(k) = mode(j)$. Note that if both $mode(i)$ and $mode(j)$ are defined, it follows from the MCDF construction rules that $mode(i)$ must be equal to $mode(j)$.

Definition 6.11 (Edge Modality). The mode of an edge $(i, j) \in V$ is represented by the function $emode : E \to \{1 \ldots M\}$, which is given by

$$emode(i, j) = \begin{cases} mode(i) & \text{if } mode(j) = \bot \\ mode(j) & \text{otherwise} \end{cases} \tag{6.3}$$

An edge is said to be **modal** if $emode(i, j) \neq \bot$. The edge is said to **belong** to mode m if $emode(i, j) = m$. The edge is said to be **amodal** if $emode(i, j) = \bot$.

6.6.1 Precedence Constraints

The start time $s(i, k, c)$ of actor i in iteration k of mode sequence c is only defined for iterations where $c(k) = emode(i, j)$ or $emode(i, j) = \bot$. In all other cases $s(i, k, c) = \bot$. Every edge $(i, j) \in V$ implies that a firing of j requires the consumption from a firing of i. In SRDF, knowing the delay of the edge is enough to know which firing of j is dependent on which firing of i, because exactly one token is consumed and produced in the edge per iteration. If there are no initial tokens on the edge, i.e. $d(i, j) = 0$, then at iteration k, firing k of j must consume the value produced by firing k of i. If there are initial tokens on the edge, then the first $d(i, j)$ iterations of j will consume initial tokens and therefore iteration $d(i, j)$ (i.e. the $1 + d(i, k)$th iteration, since we are counting from 0) must consume the output produced by the iteration 0 (i.e., the first iteration) of i, the iteration $d(i, j) + 1$ of j consumes the output of iteration 1 of i, and so on. In MCDF, the same rule applies to amodal edges, since, according to Theorem 6.1, one input is produced and an output is consumed from these edges for every iteration of the MCDF graph. Therefore, the precedence constraint due to an amodal edge (i, j) is given by

$$s(j, k, c) \geq s(i, k - d(i, j), c) + t(i), d(i, j) \leq k. \tag{6.4}$$

For modal edges, however, things are different. Modal edges only have productions and consumptions on iterations where the mode controller selected their mode for execution. Take an edge (i, j) such that $(i, j) \in E$ and $emode(i, j) = m$. If such an edge has an initial number of tokens $d(i, j) > 0$ in it, the first $d(i, j)$ iterations where mode m is selected, j will not require any token produced by i, as it will be consuming one by one the initial tokens in (i, j). Say that for a specific mode sequence c, the $(d(i, j) + n)$th selection of mode m happens at iteration k. Then the firing of actor j at iteration k is dependent on the firing of i that occurred at an iteration k' where mode m was selected, such that $d(i, j)$ selections of mode m happen between $c(k')$ and $c(k)$. This is given by $k - \delta(k, c, d(i, j))$. So, the precedence constraint due to a modal edge (i, j) is given by

$$s(j, k, c) \geq s(i, k - \delta(k, c, d(i, j)), c) + t(i) \tag{6.5}$$

when $emode(i, j) = c(k) \wedge \delta(k, c, d(i, j)) \leq k$.

6.6.2 Self-Timed Execution

As for SRDF, a self-timed schedule of an MCDF graph is a schedule where all firings of all actors happen as soon as possible, taking into account their precedence constraints. For MCDF, however, a different worst case self-timed schedule exists for each different mode sequence. The start time for the firing of actor $j \in V$ on iteration $k \geq 0$ for a mode sequence c, is defined only if $mode(j) = c(k) \vee mode(j) = \perp$. In this case the start time function is given by:

$$s_{wcsts}(j,k,c) = \max_{(i,j)\in V} \begin{cases} s_{wcsts}(i, k - \delta(k,c,d(i,j)), c) + t(i), \\ \quad \text{if } k \geq \delta(k,c,d(i,j)) \wedge emode(i,j) = c(k) \\ \\ s_{wcsts}(i, k - d(i,j), c) + t(i), \\ \quad \text{if } k \geq d(i,j) \wedge emode(i,j) = \perp \\ \\ 0, \\ \quad \text{otherwise} \end{cases} \tag{6.6}$$

If $mode(j) \neq c(k)$ and $mode(j) \neq \perp$, then $s_{wcsts}(j,k,c) = \perp$.

6.6.3 Rate-Bounding SRDF

In this section we will show that a conservative, periodic bound exists on the times of all firings of an MCDF graph that is given by worst-case self-timed schedule of an SRDF graph obtained by replacing all conditional productions/consumptions by unconditional ones.

Theorem 6.3. *Let $G = (V, G, t, d, M, amode, atype)$ be an arbitrary, well-constructed MCDF graph, an let $G' = (V, G, t, d)$ be an SRDF graph obtained by taking all actors and arcs from G, that we refer to as the* ***rate-bounding SRDF graph*** *of G and their respective timing and delay annotation, but where all actors are assumed to have fixed production and consumption rates of 1 per firing on all edges. Then, for any actor i, and for any mode sequence c and iteration k such that $c(k) = mode(i) \vee mode(i) = \perp$, it holds that*

$$s_{wcsts}(i,k,c) \leq s'_{wcsts}(i,k) \tag{6.7}$$

where $s'_{wcsts}(i,k)$ is the worst-case self-timed schedule of graph G' as defined in Chap. 4.

Proof. Recall from Chap. 4 that the WCSTS of SRDF graph G' is given by:

$$s'_{wcsts}(j,k) = \max_{(i,j)\in V} \begin{cases} s'_{wcsts}(i,k-d(i,j)) + t(i), & \text{if } k \geq d(i,j) \\ 0, & \text{otherwise} \end{cases} \tag{6.8}$$

We will prove the theorem by induction on the length of the chain of dependencies.

1. Base step: We must prove that for actor firings (i,k) (i.e. the firing of actor i at iteration k) that are not dependent on other actor firings, it holds that $s_{wcsts}(i,k,c) \geq s'_{wcsts}(i,k)$. This happens for actors that have non-delay-less input edges and for iterations where these actors can fire by consuming initial tokens only. For amodal actors, one can see by inspection of the formulas for s_{wcsts} and s'_{wcsts}, that the start time of such firings happens at time 0 for both graphs and schedules, thus upholding the desired property.
   For modal actors, if the firing (i,k) occurs and is independent of other firings, its start time for the MCDF schedule is $s_{wcsts}(i,k,c) = 0$. The (i,k) firing for SRDF graph G' always exists and must be either 0 or positive $s'_{wcsts}(i,k) \geq 0$, therefore upholding the desired property.
2. Induction step: for any actor firing (i,k) that is dependent on previous firings, we must prove that if, for all the firings (j,k') that it depends on, it holds that $s(j,k,c) \leq s'(j,k')$, then it follows that $s(i,k,c) \leq s'(i,k)$.

The right-hand side of the equation that defines s_{wcsts}, as given by Eq. 6.6 is a *max* expression on all the predecessor edges of j. The contribution to the *max* expression of each predecessor edge is dependent on whether the edge is modal or amodal, an on whether the firing count k is high enough that iteration k of j is not consuming initial tokens on arc (i,j). We will show that none of these contributions can cause $s_{wcsts}(j,k',c) > s'_{wcsts}(j,k)$ when the inductive assumption $s(j,k',c) \leq s'(j,k')$ holds.

If for an edge (i,j) and iteration k, the contribution of the edge to the *max* expression is given by the otherwise branch, then that contribution is 0. Since scheduling times for SRDF graphs are always positive or 0 (which can be easily verified by inspection of Eq. 6.8, this contribution cannot cause $s_{wcsts}(j,k,c) > s'_{wcsts}(j,k)$.

For amodal edges (i,j) and iterations k such that $k > d(i,j)$, it is simple to see that if our induction hypothesis holds, i.e. $s_{wcsts}(i,k-d(i,j),c) \leq s'_{wcsts}(i,k-d(i,j))$, then the contribution of this edge to $s_{wcsts}(j,k,c)$, given by $s_{wcsts}(i,k-d(i,j),c)+t(i)$ is upper-bounded by $s'_{wcsts}(i,k-d(i,j))+t(i)$. Therefore, if one such contribution is dominant, then $s_{wcsts}(j,k,c) \leq s'_{wcsts}(j,k)$.

For modal edges (i,j) and iterations k, such that at $k > \delta(k,c,d(i,j))$, the contribution to the *max* expression in Eq. 6.6 is given by $s_{wcsts}(i,k-\delta(k,c,d(i,j)),c)+t(i)$. Because of our induction hypothesis, this contribution to the max expression is upper-bounded by $s'_{wcsts}(i,k-\delta(k,c,d(i,j)),c)+t(i)$. For the same edge$(i,j)$ and iteration k, there is a contribution to the *max* expression that defines $s'_{wcsts}(j,k,c)$, as given by Eq. 6.8, that is given by $s'_{wcsts}(i,k - d(i,j)) + t(i)$. Now, due to Lemma 6.1, it always holds that $\delta(k,c,d(i,j)) \leq d(i,j) \Leftrightarrow k - \delta(k,c,d(i,j)) \leq k - d(i,j)$. Due to FIFO ordering of SRDF, $s'_{wcsts}(i,k - \delta(k,c,d(i,j))) \leq s'_{wcsts}(i,k - d(i,j))$ and, therefore $s_{wcsts}(i,k - \delta(k,c,d(i,j)),c) + t(i) < s'_{wcsts}(i,k - d(i,j)) + t(i)$. Therefore, if one such contribution is dominant in the max expression of s_{wcsts}, then again it holds that $s_{wcsts}(j,k,c) \leq s'_{wcsts}(j,k)$.

We have covered all three types of terms that contribute to the *max* expression in Eq. 6.6. We conclude that no matter which of the edges defines the value of the *max* expression in Eq. 6.6, the induction holds for all edges and iterations on any well-constructed MCDF graph. □

A direct consequence of this theorem is that, if the rate-bounding SRDF graph G' has a finite MCM of $\mu(G')$, then all amodal actors in the original MCDF graph will be able to fire at least once every $\mu(G')$ time units. Modal actors will be able to do the same as long as the mode to which they belong is repeatedly selected for consecutive iterations.

6.6.4 Mode-Sequence Specific Reference Schedule

In the previous section we saw how a periodic bound that is guaranteed to be met by all modes can be obtained by applying SRDF analysis to the MCDF graph as a whole. Although such analysis may be sufficient if our intention is to guarantee that, no matter what mode is executed for a particular iteration or how transitions between modes occur, the graph still meets an overall throughput requirement, this may not be enough for all intended usages of MCDF. One of the problems is that such analysis doesn't take into account that each mode has a different latency from source to sink. If, as in the case of the WLAN, we are interested in bounding the latency of a particular sequence of mode executions, this method will grossly overestimate this latency, since it will represent every iteration as taking the time of the mode with the longest latency. Also, in some cases it may be acceptable that the execution of a certain mode cannot really meet the minimum period requirement of the source, when such mode is only executed for a small finite number amount of times in a mode sequence of interest (fort instances, the mode sequence necessary to decode a complete WLAN packet), and the graph has a recovery time after such a sequence where it can flush the input buffer and catch-up with the source. This is actually the case for the WLAN: after a complete packet is received, the sender goes on a hiatus that allows the receiver to recover any extra latency it incurred by executing a number of iterations at a rate slower than the source. In such a case, we are interested in calculating a bounded buffer that can handle the largest mode sequence of interest, and provide a certain bound to the total latency of executing the sequence, but different modes may have different periods.

Moreover a tight bound on the latency of data flow execution for a particular mode sequence can be obtained by resorting to worst-case self-timed simulation of the MCDF graph for a specific mode sequence. This method will yield tighter performance results than the method of computing the MCM of the MCM-equivalent SRDF graph, since the latter method essentially results in the equivalent of assuming that the mode with the slowest initiation interval is selected for all iterations. Since MCDF graphs are monotonic, the firing times in any execution of the graph with maximum action urgency (i.e., self-timed), but with potentially varying execution times, for a given mode sequence of interest are guaranteed to be earlier than in the worst-case self-timed simulation.

Although this simulation method works for many practical cases where the number and length of all the mode sequences of interest are relatively low, in this section we will present a third technique that can be used to quickly obtain a mode-sequence-specific upper bound on the firing times of actors on an MCDF graph.

We start by observing that, if an MCDF graph G is executing for a number of iterations in the same mode m, this is equivalent to executing for those iterations the mode m subgraph G_m of G.

This SRDF graph G_m has its own MCM, say μ_m. Therefore, every actor in it can fire periodically with a minimum period given by μ_m, and at least one rate-optimal SPS can be built that bounds the firing times of all actors in any of its self-timed schedules, as shown in Chap. 4. We will call one such SPS a **reference schedule** for mode m, denoted as s_m, where $s_m(i)$ yields the firing time of the first iteration of actor $i \in G_m$. In Chap. 4, we always assumed that such a schedule would start executing from time 0, but if in any such schedule we shifted all execution times by adding a positive constant, we would also have a valid SPS schedule of the SRDF graph, starting from the initial token distribution. This is easy to verify by inspecting the precedence constraints (add the same constant to $s_{SPS}(i)$ for all $i \in G$ and all precedence constraints are preserved). Incidentally, this also means that we can always obtain an SPS schedule of a graph where there is at least one actor that starts at time 0 (take any SPS schedule of the graph, determine which actor has the lowest start time and subtract that value from the start time of all actors in the SPS to obtain a new SPS where at least one actor starts at time 0). We will assume that all SPS schedules we generate have at least one actor i such that $s(i) = 0$ (and no actor j has $s(j) < 0$).

Now say that we want to provide a bound for the firings at the k iteration of graph G, and that we know that mode m will be selected for that iteration. We will first assume that we only need to care about the dependencies for the previous iteration. Later, we will take into account dependencies for the previous iteration of the current mode. Assume that we also know all the finish times (or conservative bounds for all finish times) of actor firings in the previous iteration, $k - 1$. The execution of G_m at iteration k will have dependencies from previous iterations. These will be due to arcs $(i, j) \in G$, such that $d(i, j) > 0$, $j \in V_{m_1}$. To simplify the problem, we will consider that, if $d(i, j) > 0$, then $d(i, j) = 1$. This is because any MCDF graph where $d(i, j) > 1$ can be converted onto a graph where $d(i, j) \leq 1$ with the same timing constraints, by replacing any arc with $d(i, j) > 1$, by inserting $d(i, j) - 1$

new SRDF actors $\{a_1, a_2, \dots, a_{d(i,j)-1}\}$, each with an execution time of 0, and edges $\{(i, a_1), (a_1, a_2), \dots, (a_{d(i,j)-1}, j)\}$ each with a delay 1, for a total delay of $d(i, j)$ in the path between i and j. By assuming that we make this transformation before analyzing the temporal behavior of the graph, we need only to look at the dependency of the current iteration to the previous one.

Say that we shift all starting times in this schedule forward in time by a positive constant K, such that all start times in the SPS schedule are guaranteed to happen after all data dependencies from the previous iteration have been satisfied. We then have a valid bound for the execution time of any actor $i \in G$ at iteration k, given by $s(i) + K$. So the question is now how to chose a valid value for K.

Now, let's assume we have a similar upper bound on $s(i, k-1)$, that is, we have an SPS schedule for the mode given by $c(k-1) = m_2$, given by $s_{m_2}(i)$ for all $i \in V_{m_2}$, and we know that constant K' meets equivalent requirements with respect to all $s(i, k-1)$ as K meets for $s(i, k)$.

Since $s_{m_2}(i)$ is an SPS schedule with MCM μ_{m_2} we know that any actor in this schedule is ready to execute again after μ_{m_2} time units after the previous one, therefore, for valid K, K', and for every amodal actor in the MCDF graph it must hold that

$$\begin{aligned} s_{m_1}(i) + K \geq s_{m_2}(i) + K' + \mu_{m_2} \Leftrightarrow \\ K - K' \geq s_{m_2}(i) - s_{m_1}(i) + \mu_{m_2} \end{aligned} \tag{6.9}$$

What is interesting about this expression is that it provides a difference between a valid K bound and a valid K' bound for the transition between modes m_1 and m_2 that is only dependent on the SPSs we chose as reference schedules for mode m_1 and m_2, and not on the particular iteration at which the transition happens. Given that we know K', i.e. a time shift that can be added to s_2 in order to provide a valid bound for the firing of actors in mode m_2 on a particular iteration of the MCDF graph for a particular mode sequence, we can compute a bound for the subsequent iteration. We will call the difference $K - K'$, the **mode transition interval** $\kappa_{m_2 m_1}$, from mode m_2 to mode m_1, defined by:

$$\kappa_{m_2 m_1} = \max_{i \in G_{m_2} \cap G_{m_1}} (s_{m_2}(i) - s_{m_1}(i) + \mu_{m_2}) \tag{6.10}$$

Notice that Eq. 6.10 implies that $\kappa_{m_2 m_1} \geq \mu_{m_2}$ since, because of our choice of reference SPS schedules, there must be an actor i such that $s_{m_1}(i) = 0$ and for such actor, it is true that $s_{m_2}(i) - s_{m_1}(i) + \mu_{m_2} \geq \mu_{m_2}$, as $s_{m_2}(i) \geq 0$.

What about transitions from a mode to itself? Since s_m is a SPS for G_m, we know that a second iteration in the same mode can start after an interval of time equal to $\mu(G_m)$. And in fact, Eq. 6.10 gives us $\kappa_{mm} = \mu(G_m)$.

We have shown how a static periodic schedule can be delayed to account for inter-mode dependencies, and provide a bound for the execution of the graph with mode m at iteration k, based on a bound for the start of the execution of the previous iteration $k-1$. We know a valid starting time $K(k, c)$ for iteration k and mode sequence c must be such that

$$K(k,c) \geq K(k-1,c) + \kappa_{c(k-1)c(k)}. \quad (6.11)$$

However, we have only considered amodal actors. For modal actors, the dependency is towards the last iteration at which mode m was chosen for execution. In fact, $K(k,c)$ must be such that

$$K(k,c) \geq K(k-\delta(k,c,1),c) + \mu_{c(k)}. \quad (6.12)$$

If we chose to compute the bounds as

$$K(k,c) \geq K(k-1,c) + \kappa_{c(k-1)c(k)} \quad (6.13)$$

we comply with both constraints. This is self-evident for the constraint given by the inequality in Eq. 6.11. The second inequality is also taken into account, since the iteration that follows an iteration $k-\delta(k,c,1)$ as in Eq. 6.12 has a bound, as given by Eq. 6.13 of

$$K(k-\delta(k,c,1)+1,c) = K(k-1,c) + \kappa_{c(k-1)c(k)} \quad (6.14)$$

Since we already established when discussing Eq. 6.10 that $\kappa_{mn} \geq \mu_m$, then the previous equation implies that

$$K(k-\delta(k,c,1)+1,c) \geq K(k-\delta(k,c,1)) + \mu_m \quad (6.15)$$

Since a positive k is always greater or equal to $k-\delta(k,c,1)+1$ and $K(k,c)$ is always greater or equal to $K(k-1,c)$ according to Eq. 6.13, then any $K(k,c)$ as given by Eq. 6.13 respects the constraint given by Eq. 6.12.

Having computed the upper bound $K(k,c)$, the start time of a given actor i, amodal, or of mode $c(k) = m$, at iteration k can be bounded by $s(i,k,c) \leq K(k,c) + s_m(i)$, where s_m is a chosen static periodic schedule of mode m subgraph G_m. This provides a way of computing upper bounds for all start times in the self-timed schedule by computing the MCMs of all mode subgraphs, one static periodic schedule per mode graph, and the κ_{mn} mode transition intervals for all mode transitions from mode m to mode n.

This method can, for long mode sequences, be much more efficient than data flow simulation. Also, it allows for the fast temporal analysis of multiple mode sequences.

6.6.5 Temporal Analysis of Variable-Length Mode Sequences

The technique described in the previous section can be extended to handle mode sequences of variable length. Suppose we are interested in analyzing the temporal behavior of a family of mode sequences where a mode transition sequence is given by $\vec{m} = [m_1, m_2, \ldots, m_n]$, but where each mode is executed for a number of times before each mode changes. Let $\vec{p} = [p_1, p_2, \ldots, p_n]$ such that m_1 is executed for $p_1 > 0$ iterations, mode m_2 is executed for $p_2 > 0$ iterations and so on. Any two subsequent modes in the sequence of mode transitions is different, i.e. $m_q \neq m_{q+1}$ for all $q < n$. Let the family of mode transitions defined in this way be denoted by C, and $C(\vec{m}, \vec{p})$ is the mode sequence for the mode transition sequence $\vec{m}$ with the number of executions by transition given by $\vec{p}$.

We are interested in bounding the latency of $s(i, k, C(\vec{m}, \vec{p}))$, where $k = \sum_{q=0}^{n} p_q$, and $i \in V_{m_n}$. We do this by summing up the values of $K(k, C(\vec{m}, \vec{p}))$ for all transitions. However, since we do not know the exact number of iterations that are executed between two transitions, we have to conservatively approximate the value of the second term of the max expression in Eq. 6.13. Since we assume that all $p_q > 0$, we know that at least all mode transitions have occurred between the mode of the current iteration and the previous selection of that same mode, although we do not know how many iterations were executed in each mode. We can then infer that the distance between two consecutive iterations of the same mode has to be at least equal to the number of modes visited between the two. This distance is equivalent to the modal distance for a mode sequence where each mode in the mode transition sequence executes exactly once, that is $C(\vec{m}, \vec{1})$, where $\vec{1} = [1, 1, \ldots, 1]$. This modal distance is given by $\delta(k, C(\vec{m}, \vec{1}), 1)$ and therefore a conservative approximation of $s(i, k, C(\vec{m}, \vec{p}))$ is given by

$$\begin{aligned} s(i, k, C(\vec{m}, \vec{p})) \leq\ & \mu_{m_1}.(p_1 - 1) + \kappa_{m_1 m_2} \\ & + \mu_{m_2}.(p_2 - 1) + \kappa_{m_2 m_3} \\ & + \ldots \\ & + \mu_{m_n}.(p_n - 1) + \kappa_{m_{n-1} m_n} \\ & + s_{m_n}(i). \end{aligned} \tag{6.16}$$

This can be factored into:

$$s(i, k, C(\vec{m}, \vec{p})) \leq \sum_{q=0}^{n} \mu_{m_q}.(p_q - 1) + \sum_{q=1}^{n} \kappa_{m_{q-1} m_q} + s_{m_n}(i) \tag{6.17}$$

We can now analyze the graph for temporal requirements of the type

$$s(i, k, C(\vec{m}, \vec{p})) \leq L(\vec{m}, \vec{p}), \tag{6.18}$$

where L is a function that provides a temporal bound that is parameterized on the number of repetitions of each mode in the family of mode transitions $C(\vec{m}, \vec{p})$.

6.6.6 *Temporal Analysis Summary*

We proposed four different techniques to perform worst-case temporal analysis of MCDF graphs:

- Bound on overall throughput and start times of all actor firings as given by MCM analysis and SPS scheduling of a rate-bounding SRDF graph;
- Worst-case Data Flow simulation of all mode sequences of interest;
- Generation of reference schedules per mode and mode transition intervals, to be composed through Eq. 6.13 for mode sequences of interest; note also that the worst case mode transition interval provides an upper bound to the period of any mode sequence;
- Generation of symbolic expressions on the number of executions of each particular mode, for parameterized mode sequences of interest.

Each technique has its own advantages and limitations. The first technique is simple, as it re-uses without significant changes the techniques used for SRDF analysis. The problem with this technique is that it always takes the worst-case throughput across all modes as the worst-case throughput of the graph when executing in any mode. The second technique allows the most accurate bounds, but it requires a complete data-flow simulation of the MCDF graph for all mode sequences of interest, which may be too costly in terms of computation time if many and/or very long sequences need to be considered, not to mention the fact that it cannot cope with virtually infinite sequences, at least not in the general case. The third technique reduces the accuracy of the analysis, since it resorts to static periodic schedules as bounds on the temporal behavior of the self-timed execution, but allows much faster computation of bounds per sequence. The fourth technique is an extension of the third, and allows the analysis of sequences of variable length. It also provides, for virtually infinitely repeating sequences, a general expression for the bounds on firing times.

One further elaboration of the two last analysis techniques is possible. We can generate all the reference SPS as a single Linear Program formulation, including the expressions for the mode transition intervals, and minimize the weighted sum of the mode transition intervals. The weight of each mode transition interval is chosen such that the most common transitions (according to the mode sequences of interest) have heavier weights in the objective function.

6.7 Scheduling MCDF Graphs

We will compute schedules by extending to MCDF the techniques we proposed for SRDF schedules in Chap. 5. With that in mind, we will proceed to explain how our techniques to generate static-order schedules, budgets for Run-Time Schedulers (RTSs) and sufficient buffer sizes can be extended for MCDF.

6.7.1 Generating Quasi-Static-Order Schedules

To extend the scheduling framework described in Chap. 5 so that it accommodates for MCDF graphs, we need to make a number of changes. First, the data-dependent nature of the MCDF execution model does not allow for static-order schedules. A body of work exists in creating a schedule as close as possible to static by leaving only data-dependent decisions to be performed at run-time. This is referred to as quasi-static scheduling [10, 37]. In quasi-static scheduling, the start times of all actors are still fixed, as in a static schedule, but depending on a data value, the actor is either fired or not at the pre-determined time, that is, we decide at run-time whether the actor fires or not, but, if so, the time of the firing was chosen at compile time. For reasons already explained in Chap. 2, we do not want to fix the start times of all actors, but rather use a self-timed approach to actor firing, where actor ordering plays the role of reducing the uncertainty caused by local arbitration. Because of this, our local scheduling of each mode is not static, but statically-ordered, and thus we call our local, compile-time generated, intra-job scheduling policy Quasi-Static Ordering (QSO), i.e., a schedule that is as close as possible to statically ordered, where only the data-dependent decision on what mode to execute is left to be performed at run-time.

A QSO is very similar to a static ordering. For each MCDF iteration, all actors mapped to a given processor are given the right to fire in a sequence predefined at compile-time. When an actor is given the chance to fire, it will wait for its input data dependencies, as specified by the DF input graph to be met (also, output space dependencies for bounded output buffers must be met). Once the firing conditions are met, the actor fires, and the next actor in the QSO is given the right to fire. This is the same as static-ordering, except that in QSO, modal actors whose mode was not selected for execution in this iteration are skipped.

Such execution model assumes that the local scheduler is aware of the mode control signal by the time it decides to schedule the first modal actor in the QSO sequence.

Because of this, we must change the DF scheduler we presented in Chap. 5, in that the DF now automatically inserts in every processor's static order a newly created Mode Switch that is responsible for splitting the schedule in modal branches. This Mode Switch is inserted just before the first modal actor is inserted in the QSO, and it must be connected to the control output of the Mode Controller, just like any Mode Switch in an MCDF graph. We will refer to such a Mode Switch as the Processor Switch of a given processor for a given QSO of an MCDF graph.

Since, according to the construction rules, in any valid MCDF any modal actor on a given iteration can be scheduled after the MC, any QSO thus generated results in a valid schedule.

Therefore, our DF scheduler can, with one minor change, generate valid QSO schedules. An issue that we must still address is how to model the temporal behavior of the execution of the MCDF graph, subjected to the generated QSO.

6.7.2 Modeling Quasi-Static Order Schedules in MCDF

As we have seen for static order schedules in Chap. 5, the static ordering of amodal actors can be represented in the temporal analysis model by inserting an edge from each actor to its direct successor. Between modal actors of the same mode, the same modeling construct patently holds, that is, we insert an edge between any modal actor and the next modal actor of the same mode in the QSO. Between modal actors of different modes, no such dependency exists, since at each iteration actors of modes which have not been selected are skipped (i.e., they do not even check for firing conditions). Also, for each mode and each processor, the first modal actor in the QSO must be connected to the Processor Switch that was inserted during generation of the QSO.

If the QSO ends with a modal actor, we add a select to join all branches and we insert an edge with one delay between the last actor and the first actor in the QSO, to represent the no-overlapping, one-iteration-at-a-time behavior of the QSO.

What if the QSO imposed precedence constraint from a modal actor to an amodal actor? Such dependency can happen, since the DF scheduler is free to place an amodal actor i to follow a modal actor j in the QSO. Note that such dependency in the order of execution is only valid while executing an iteration for which the mode of j is selected for execution. For any other mode, i will have to wait for the last previous actor that belongs to that particular mode, or is amodal.

To better illustrate the problem assume that the QSO ordering for a given processor p is $qso(p) = [A, B, C, D, E]$, where A, B, C, D, E are actors on an MCDF graph, and $m(A) = m(B) = m(D) = \perp$, $m(C) = 1$ and $m(E) = 2$. Furthermore, B is the Processor Switch of processor p. The meaning of the QSO is that, for an iteration where mode 1 is selected by the mode controller, the execution order is A, followed by B, followed by C, followed by D; for an iteration where mode 2 is selected, the execution order is A, B, D, E; for all other iterations, the order is A,a_1,a_4.

To represent the conditional dependency of amodal actor firings on modal actor firings, we have two options:

1. **Interpolation of branches and merges**. We fix the QSO by inserting a Select before every amodal actor that appears right after a modal actor in the QSO, except if that actor is a Select, since selects can merge the schedule. Conversely, every time a modal actor appears after an amodal actor in the QSO, we insert before it a Switch to branch the schedule in modes, except if the actor is a Switch, since switches can branch the schedule. Figure 6.6a depicts this way of modeling QSO for the previously described example. We omitted the edges from the MC for simplicity sake.

   The Processor Switch is the first of the Switches. It marks the point from which the execution of the iteration is dependent on the output of the MC.
2. **"Modalization" of amodal actors**. Any amodal actor that appears after a Processor Switch in the QSO is replaced by its set of modal copies, one for

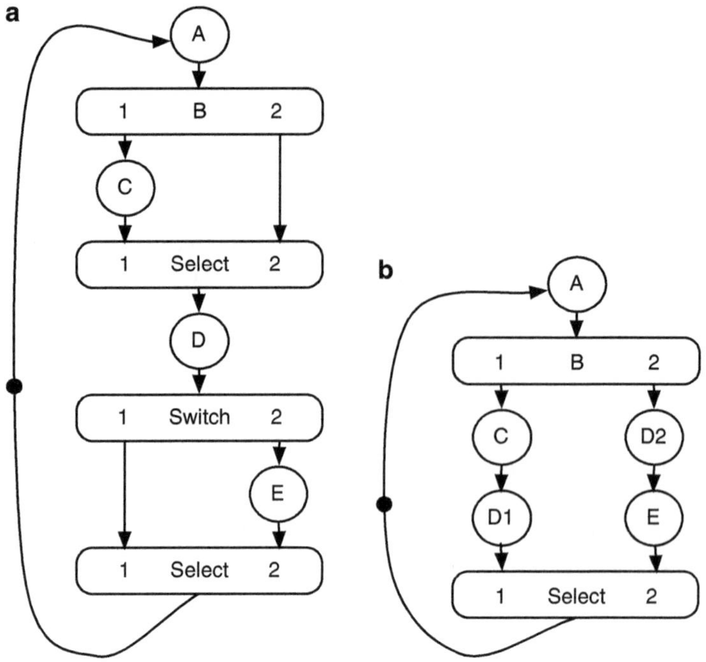

Fig. 6.6 Two options for modeling quasi-static ordering in MCDF

each mode. Figure 6.6b depict this way of modeling QSO for the previously described example. D1 and D2 are the copies of actor D, where $m(D1) = 1$ and $m(D2) = 2$.

The following procedure can be used to convert an amodal actor to a set of modal actors. A copy of the amodal actor is created for every mode in the MCDF graph. Then, any incoming edge of the original amodal actor is connected to the data input port of a newly created Switch. Each output of this Switch is connected to t the input of the copy of the amodal actor associated with the mode of that port. Any outgoing edge of the original amodal actor is connected to the output port of a newly created Select. Each input port of that select is connected to the output of the copy of the amodal actor associated with the mode of that port. Figure 6.7 depicts how an amodal actor B with one input edge and one output edge can be converted to a modal representation on an MCDF graph with two modes. Please note that this conversion makes each firing of B data-dependent on the firing of the MC on the same iteration due to the control input of the Switch. The MC and the edges from MC to the switch and select actors are not depicted.

The conversion of an amodal actor to a modal representation is going to cause the creation of many new switch and select actors and their associated edges in the analysis model.

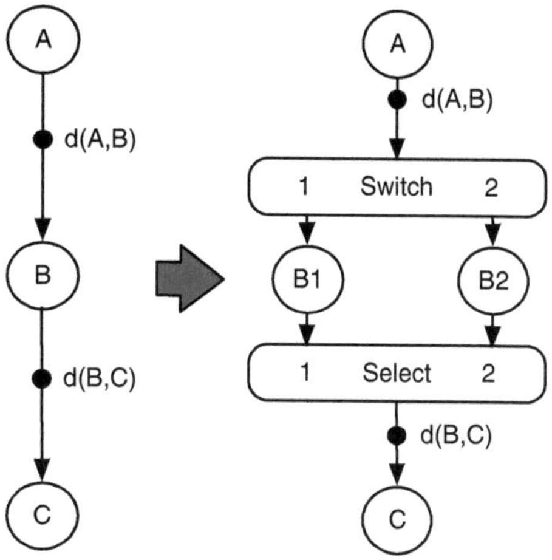

Fig. 6.7 An amodal actor is converted to a modal representation

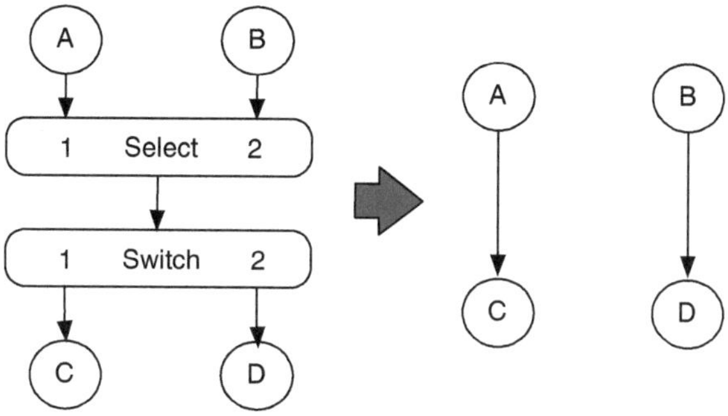

Fig. 6.8 Needless "Select followed by Switch" constructs can be removed to simplify the analysis graph

However, many of these can be removed through two simplification techniques: join-followed-by-branch removal and tunnel removal.

Join-followed-by-branch removal removes a switch followed by a select, when the edge between them has no initial tokens in it. This is depicted in Fig. 6.8.

Tunnel removal removes from the temporal analysis graph tunnel constructs (see Sect. 6.2.3) between modal actors that are mapped in the same processor. This is possible because if the order of execution between two actors of different modes executing on the same processor is already defined by the combination of the mode sequence and the QSO, that is, it is guaranteed that the producer will always execute before the consumer, since, within a processor, there is no iteration overlap.

Figure 6.9a depicts an example MCDF graph, and Fig. 6.9b shows its possible QSO schedule on a two processors system derived by applying the strategy we describe above. The dashed edges indicate the extra inter-actor dependencies generated by the quasi-static ordering.

For instance, if in the QSO of a given processor, actor a succeeds to actor b, then an edge is added from b to a in the analysis graph. When a Switch actor is added to the schedule in order to split mode-specific branches, it is also added to the analysis graph.

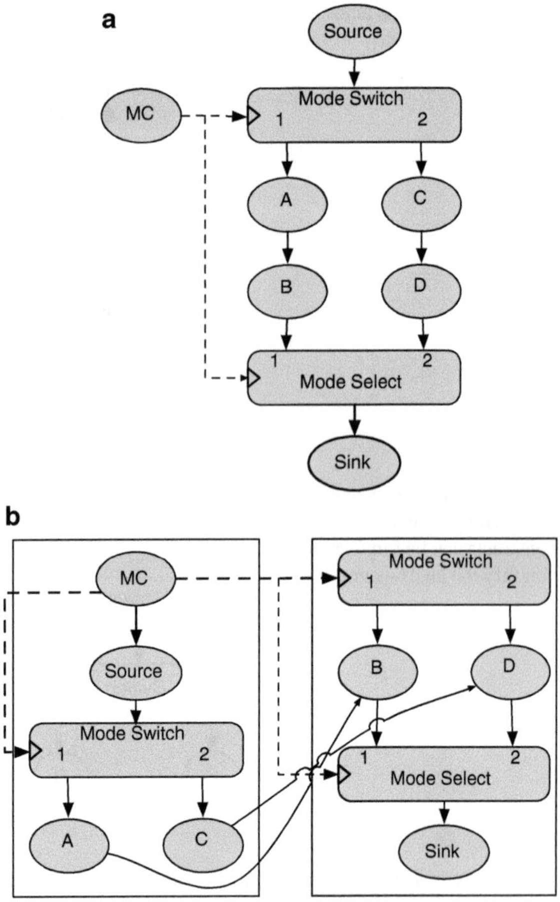

Fig. 6.9 An MCDF graph and its quasi-static order schedule on two processors

As in the original backtracking scheduler presented in [75], each ordering decision triggers a test to see if there was no infringement of temporal requirements, by using one of the 4 analysis techniques for MCDF that were previously presented in this chapter.

6.7.3 Determination of Run-Time Scheduler Settings

Run-time scheduler settings can be modeled in a similar way as we described for SRDF. If the RTS is a TDM scheduler, then we can simply compute the execution of the actors in the model exactly as in SRDF. For a round-robin scheduler, it is necessary to remember that the size of the slice is equal to the maximum sum of execution times across all modes, that is

$$S(p) = \max_{m \in M} \sum_{\substack{i \in V_m \\ \pi(i)=p}} t(i). \tag{6.19}$$

The determination of the RTS settings can be performed by using either the Binary Search Slice Allocator or the Random Slice Allocator, where at each step the temporal analysis is performed using one of the four methods described in this chapter.

6.7.4 Determination of Buffer Sizes

With respect to buffer sizing, [10] proved that the problem of finding a schedule that guarantees execution in bounded buffer size is, in general, undecided for a BDF. The same does not apply to our model.

As we have seen in Sect. 6.6.3, for any MCDF graph we can generate a rate-bounding SRDF graph. Since this SRDF can be executed in bounded buffer space [80], and there is one firing on this graph for each token produced and consumed in the original MCDF graph, we know that we can implement our MCDF graphs in bounded buffer space.

In this section, we willprovide bounds for buffer sizes for both modal and amodal edges of MCDF graphs.

In the case of modal edges, a bound can easily be computed, since consumptions and productions in these edges only occur when their mode is selected for execution. As such, the producer can never fire at a rate higher than μ_m, if $\mathrm{emode}(i, j) = m$. Therefore, we can simply compute the bound for the SPS schedule of G_m, as if for an SRDF graph:

$$b(i, j) \geq \frac{s_m(j) + t(j) - s_m(i) + \mu_m.d(i, j)}{\mu_m} \tag{6.20}$$

where $\mathrm{emode}(i, j) = m$.

As for amodal edges, the story is a bit more complicated. A bound to the required buffer size may be derived from our knowledge that the producer can never execute faster than the lowest MCM amongst the MCMs of all mode sub-graphs, say $\check{\mu}(G) = min_{m \in M} \mu(G_m)$. For arcs with 0-delay, the bound on the buffer size becomes

$$b(i, j) \geq \frac{s_m(j) + t(j) - s_m(i)}{\check{\mu}(G)} \tag{6.21}$$

for all modes $m \in M$ such that $emode(i, j) = m$ or $emode(i, j) = \perp$. If $d(i, j) \neq 0$, then we again have the same problem we encountered with respect to temporal analysis, i.e., that the execution of a previous iteration of an actor is subject to the dependencies of a different mode sub-graph and, therefore to a different rate. In that case, we use the same expedient as before, where we convert the input graph such that for all edges the delay is one or 0, and we now have to consider as a bound on the buffer size the maximum bound amongst all the possible mode transitions

$$b(i,j) \geq \max_{\substack{m \in M \\ n \in M}} \left(\frac{s_m(j) + \kappa_{nm} + t(j) - s_n(i)}{\check{\mu}(G)} \right) \tag{6.22}$$

We can now, given an MCDF graph, compute sufficient buffer size by simply computing first the MCMs for all mode sub-graphs, then all SPSs for all the mode sub-graphs, then all the transition intervals, then all the bounds for the buffer sizes of all arcs, and then, per arc, take the lowest integer value that respects those bounds.

6.8 Scheduling Experiments

We applied the MCDF scheduling flow to the WLAN receiver depicted in Fig. 6.5, mapped to the same heterogeneous multiprocessor we used as a platform in Chap. 5. We ran the MCDF graph of the WLAN through the scheduling flow for two different mode sequences: one for the shortest possible packet, i.e. a mode sequence where the synchronization mode runs twice, the header mode once, and the payload mode once, and another for the longest packet, i.e. the same as the previous, except that the payload mode runs 256 times. We used both the SPS-based analysis and simulation-based analysis for constraint verification. The results are presented in Tables 6.1–6.4. In these tables, the highest priority is 3 and the lowest priority is 1. We expected to find that the overestimation of the execution time of the sequence with SPS-based analysis would cause final slice sizes to be larger that for simulation-based analysis, and that was indeed the case. We expected the worst overestimation to occur for the small sequence, as the main overestimation is caused at the start of execution of a mode, and during transitions, and the proportional contribution on the small sequence of transitions is much larger than for the long sequence, that stays in the payload mode for 256 consecutive iterations. This was verified by the experiment. The largest difference in required resource utilization after scheduling and slice size determination was found for the short sequence, where, for the binary search slicer algorithm we give the highest priority to minimizing the

Table 6.1 WLAN scheduling results with SPS-based analysis, mode sequence with 2× synchronization, 1× header, 1× payload, 1× crc

Priority			Reservation (%)		
EVP	SwC	ARM	EVP	SwC	ARM
3	X	X	29	45	40
X	3	X	45	19	45
1	3	3	40	45	9
1	2	3	45	39	9

Table 6.2 WLAN scheduling results with simulation-based analysis, mode sequence with 2× synchronization, 1× header, 1× payload, 1× crc

Priority			Reservation (%)		
EVP	SwC	ARM	EVP	SwC	ARM
3	2	1	23	43	43
3	1	2	23	45	39
2	3	1	41	19	43
1	3	2	45	19	34
2	1	3	41	45	8
1	2	3	45	41	8

Table 6.3 WLAN scheduling results with SPS-based analysis, mode sequence with 2× synchronize, 1× header, 256× payload, 1× crc

Priority			Reservation (%)		
EVP	SwC	ARM	EVP	SwC	ARM
1	X	X	29	45	40
X	1	X	45	29	36
2	3	1	40	44	9
4	2	1	45	39	9

Table 6.4 WLAN scheduling results with simulation-based analysis, mode sequence with 2× synchronization, 1× header, 256× payload, 1× crc

Priority			Reservation (%)		
EVP	SwC	ARM	EVP	SwC	ARM
3	X	X	27	45	45
2	3	1	44	29	45
1	3	2	45	29	35
2	1	3	35	45	7
1	2	3	45	36	7

EVP utilization. In this case, with SPS-based analysis, the utilization of the EVP resources was 29%, as opposed to 23% utilization with simulation-based analysis, i.e. an increase in the slice size of 26%. With the long sequence, as expected, the overestimations are much lower, with the maximum increase of slice size being again for the cases where we give highest priority in the slicer to minimizing EVP slices, for a total increase in the slice of 7%. It was also noticeable that, the longer the mode sequence is, the slower the running time of the scheduler when using simulation-based analysis, while it remains constant with SPS-based analysis. For this particular example, for the longer sequence, the running time of the data flow scheduler was about one order of magnitude larger than for SPS-based analysis, going from less than a minute to around 10 min.

6.9 Multi-rate MCDF

In this chapter, we discussed only graphs where data-independent actors always produce and consume one token in all their output/input arcs. However, extending MCDF to multi-rate is relatively simple, with some constraints. Take a graph where data-independent actors, except for the mode controller, are allowed to have rates of production/consumption on their ports higher than one. We compute repetitions vectors for the balance equations as described in Eq. 6.2, per each modal subgraph. If we obtain solutions such that the MC always executes once (and therefore so do all Mode Switches and Mode Selects since, due to the control edges, there must be a 1 to 1 ratio of firings between MC and Mode Switches and Mode Selects) per iteration for all modal subgraphs, then there is still a number of firings for each actor per each mode selection (i.e. per each firing of the MC) that returns the graph to the initial state, and therefore the graph can still execute for one mode selection, return to the initial state, then execute for another mode, return to the initial state, and so on. All our analysis and mapping techniques will still work by converting the Multi-rate modal subgraph into a Single-rate graph, as we have done for MRDF graphs.

It is more difficult, however, to handle graphs where repetitions vectors exist for all modal subgraphs, but the number of firings of MC is different than one for one or more modal subgraphs. In such cases, not all mode sequences are able to return the graph to the initial state. However, if we require the MC to behave such that for all valid mode sequences, each mode is selected a number of times equal to the number of repetitions of the MC on the repetitions vector of the modal subgraph for each mode, then the graph is still guaranteed to be able to fire in such a way as to return to the initial state after each sequence of firings of the same mode, and it is reasonable to assume that all analysis and mapping techniques can still be adapted to handle such graphs. This however, should be the subject of further study.

6.10 Further Reading

We have already discussed the limitations of Static Data Flow variants and of Integer Data Flow both in Chap. 3 and in the introduction of this chapter. During the last few years, however, there have been attempts in some ways similar to ours to find a better balance between expressivity and analytical properties. In such category falls the work on Variable-Phased Data Flow (VPDF) presented in [90]. With VPDF, actors cycle through a list of execution phases determined at compile time. The variable runtime behavior comes from the fact that each phase can be fired a number of times dependent on data values that can only be known during execution. Conditional behavior can be represent since phases can execute 0 or more times depending on data values. In [90] it is shown how, by knowing the maximum number of executions per phase, one can compute buffer sizes for a VPDF graph to meet a given throughput requirement. Scheduling is assumed to be dynamic, and no techniques have been proposed for static or quasi-static ordering of actors.

Another approach is taken in the work on Scenario-Aware Data Flow (SADF), a DF model first introduced in [87]. In [22], a technique is provided for real-time analysis of SADF graphs. In SADF, an application is modeled as a collection of SDF graphs, each representing individual scenarios of behavior, and a Finite State Machine (FSM) that specifies the possible orders of scenario occurrences. The paper provides techniques to analyze worst-case performance of such applications. These techniques can determine the highest throughput that can be guaranteed and can determine the minimal latency that can be attained from a state space analysis. SADF is very similar to MCDF in many ways. In fact, the two models comparable expressive power and automatic conversion between the two is possible. The analysis techniques for SADF require the same information that we use for MCDF analysis: a graph per mode/scenario and a specification of scenario transitions of interests. In our case, this is done with the specification of mode sequences; in SADF, with the specification of a finite state machine for scenario transitions. The analysis technique proposed for SADF requires the exhaustive analysis of the state space of the execution of the SADF graph, as specified by the FSM. Our usage of reference schedules makes the analysis less precise, but avoids an exhaustive search space exploration, which is also a problem we have with the analysis technique where we apply data flow simulation for all mode sequences of interest.

A difference between MCDF and SADF is that SADF was primarily designed as a way of modeling behavior and not as a programming model and, because of that, the execution model is not explicit about the relation between scenario transition decisions and data flow execution, i.e., if one defines the FSM and each one of the graphs per scenario, there is still no way of knowing where the decision to go from one scenario to another was taken. Application mapping techniques for SADF were proposed in [85]. In this work, static ordering is handled by a single attempt, heuristic algorithm that assumes that the mode of execution is known from the start of the iteration at each processor. It is not clear how this constrains the options available to the scheduler. SADF analysis is likely to be more exact and flexible than MCDF analysis, but MCDF works better as a programming model and it has more mature scheduling techniques. It is certainly possible to combine the best of the two worlds: use MCDF as the model for specifying the application and carrying out quasi-static scheduling, and converting the model to SADF – which essentially means extracting the modal graphs for each mode and representing the valid mode sequences as an FSM – for the purpose of timing analysis.

6.11 Conclusions

Transceivers are mostly iterative, repetitive applications, with little data value-dependent behavior. However, the little data value-dependent behavior there is cannot easily be accommodated by the static data flow computation models. This posed a serious problem, as, until recently, only the static flavors of data flow were known to have the analytical properties that allows us to guarantee hard-real time behavior. On one side we had the dynamic data flow variants that could not be analyzed, and on the other side we had the static data flow models that we could analyze, but were not expressive enough to model the behavior of transceivers. The work presented in this chapter tries to find a compromise: it allows a small amount of data-dependent behavior and extends the known analysis techniques for static graphs to accommodate for this small amount of dynamism. The result is the Mode-Controlled Data Flow computation model. We have shown how MCDF can be analyzed in terms of minimum guaranteed throughput and maximum latency bound and we have extended the techniques for application mapping we developed in Chap. 5 so that MCDF graphs representing radio applications can be quasi-statically scheduled onto multiprocessor systems.

There is still considerable work to be done. The proposed quasi-static ordering scheduler is an adaptation of the static ordering scheduler proposed in Chap. 5, and, as such, it still makes an exhaustive search for feasible schedules, with an algorithm of exponential complexity. For buffer sizing, we have only shown that we can compute an upper bound to the requirements, but this upper bound is not optimal.

The biggest challenge, however, is probably the expressivity of the model. The expressivity of MCDF was chosen such that the transceivers that were known to us could be conveniently expressed. There is a trend, however, for baseband processing to become more irregular, with more complex control, as can be seen in the evolution of 4G standards such as LTE. It is yet to be seen whether MCDF can cope with the increase of control complexity in baseband processing. A study has been carried out to model LTE in MCDF and MCDF os certainly capable to express all the required behavior, but the complexity of the data flow model thus obtained can be dramatically reduced by the usage of additional higher-level constructs (similar to the Mode Tunnel) that, although reducible to a sub-graph composed of Mode Switches, Mode Selects and SRDF actors and edge, will make the specification simpler to write, understand and manage.

Chapter 7
Resource Manager

In this chapter, we discuss the algorithms that the resource manager must implement to try to find a mapping for a transceiver instance (a job) at runtime, and decide whether the job can be executed or not.

In the previous chapters, we have seen how a transceiver described by a data flow graph can be prepared at compile time for runtime allocation of resources. Actors are first bundled into quasi-statically ordered clusters. Each cluster is annotated with a resource budget, which includes scheduler settings (the size of a slice per period) and memory requirements, while communication channels between clusters are annotated with required buffer capacity.

In Chap. 2, we proposed an approach to manage resources on a system with dynamic job-set and job-mix which gives jobs a degree of independence from the rest of the system. This resource management policy should ensure: *(a) admission control* – a job is only allowed to start if the system can allocate upon request the resource budget it requires to meet its timing requirements; and *(b) guaranteed resource provisions* – the access of a running job to its allocated resources cannot be denied by any other job. Resource budgets are computed offline such that jobs meet timing constraints, using the techniques guided by temporal analysis that we described in Chap. 5.

As we have argued in Chap. 2, the enforcement of resource budgets requires the hardware platform to provide predictable arbitration schemes, that is, arbitration schemes that allow us to tightly bound the time that a request takes to be served.

We designed a template for scalable multiprocessor architectures that fits these requirements. We use global resource allocation to implement admission control and local schedulers to guarantee resource provision. A simplified version of this architecture was presented in Chap. 2. In this chapter, we extend the communication infrastructure by using an Æthereal Network-on-Chip [28]. Æthereal fits well in this architecture template as it allows for resource reservation by providing connections with guaranteed minimum throughput and maximum latency. Other communication schemes can be accommodated if their worst-case temporal behavior can be conveniently approximated by a data flow model.

O. Moreira and H. Corporaal, *Scheduling Real-Time Streaming Applications onto an Embedded Multiprocessor*, Embedded Systems 24, DOI 10.1007/978-3-319-01246-9_7,

We start our discussion assuming a target platform containing a TDM-arbitrated router – i.e. an Æthereal network with a single router – for inter-tile communications. We show that the problem can be seen as a generalization of Vector Bin-Packing (VBP) [53] and, since the problem is NP-complete, provide heuristic approaches to tackle it. Since a single router architecture does not scale with an increase in the number of cores, we look at the problem assuming a network with many (4–12) routers.

By adding the network, the system changes in two important aspects. First, the resource allocator must find routes through the network and reserve time slots per link per router. Secondly, the network cannot be modeled as a single resource, as bottlenecks between two tiles are formed depending on which routes are allocated. Special care has to be taken about which processors are used for which tasks, even though the set of processors itself is homogeneous.

In Sect. 7.1, we describe the assumptions about hardware, software and scheduling mechanisms that were used in this study. In Sect. 7.2 we formally model the resource allocation problem. In Sect. 7.3, we describes the heuristics we propose to find a feasible allocation. In Sect. 7.4 we show the results of some of our experiments, as they highlight the advantages and disadvantages of the proposed algorithms. In Sect. 7.5, we explain how this work relates to the work of others. Finally, we state the conclusions for this chapter in Sect. 7.6.

7.1 System Architecture

In this section we will describe some relevant aspects of the type of system we consider. We will describe the hardware and software models, as well as the scheduling mechanisms used in the processors.

7.1.1 Hardware

In Chap. 2 we propose a hardware architecture consisting of a set of processors of different types, fully connected through a global communication framework. In this chapter, we will both simplify and extend this model. First, we will consider a homogeneous system, i.e. all processors are of the same type. The reason for this is that, since in our framework each task can only run in one specific type of processor, the problem of allocating resources to such a task is restricted to finding a processor of the type that task requires. The problem of resource allocation for our heterogeneous system can therefore be seen as jointly solving the resource allocation problem for several homogeneous multiprocessor systems.

On the other hand, in Chap. 2 we have not described in detail what the communication infrastructure looks like. If for relatively small and simple architectures a multi-layer bus may suffice (provided that care is taken to employ predictable slave-side arbiters) such a solution will not scale well with an increase in the number of cores in the architecture.

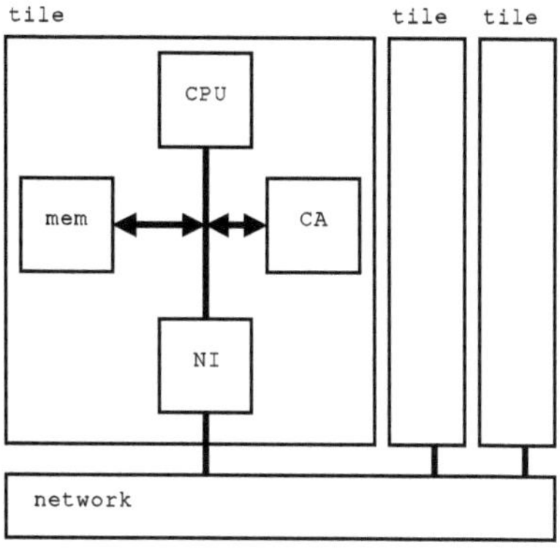

Fig. 7.1 A processing tile

In this chapter, the hardware under consideration consists of a homogeneous set of processors, connected through a network of routers. The processors and the network are placed on a single chip, forming a Multi-Processor System on Chip. We will consider at most 24 processors and 12 routers.

Each processor resides on its own *processing tile* (PT), which consists of a processor, local memory (MEM), a communication assist (CA) and a network interface (NI) (Fig. 7.1). The CA offers a fixed number of time slices that it allocates to memory accesses between the ARM and the NI. The NI offers a fixed number of input/output FIFO queues for accessing the channels allocated over the network.

For a model of the network we consider the Æthereal [28, 29] Network-on-Chip. Æthereal can use contention-free routing, based on TDM switching and allows links with guaranteed throughput and latency. Every router is connected to other routers and/or to NIs. Each PT has one single NI. Due to physical constraints, the routers can have at most eight bidirectional connections. Every connection consists of a fixed number of links, and each link offers a fixed amount of bandwidth.

Data are transmitted by the routers in three-words packets. All routers and tiles send and receive data packets synchronously as if operating under a global clock. If data arrives on an input at tick t, it will be sent through the router to the output link and arrive at the other end of that link at tick $t + 1$. Per time slice, a packet from each input link is sent to an output link, according to a slot reservation table that specifies per time slice which input link is connected to each output link.

Two processes on different PTs can communicate unidirectionally by establishing a channel. This requires a send and a receive buffer to be allocated in the data memory of their respective PTs, as well as time slices in the CA and queues in the NI on both sides. In the network, a communication channel must be created by finding a path across routers from source PT to destination PT. The routers do not buffer their input packets. Therefore, when a packet arrives to an input link in a router, it must be sent at the end of the cycle to the output link specified for that time slot in the slot reservation table.

Each processor runs either a TDM or a NPNBRR scheduler, and has a (possibly empty) set of tasks assigned to it.

7.1.2 Application

As in previous chapters, we will assume that the application consists of a number of stream-processing jobs. An unpredictable source, such as the user, can ask to start or stop a job instance at any moment. Because the user can typically request for more job instances than the hardware can run simultaneously, there is a resource allocation problem. If the system can find sufficient resources, the job is started. Otherwise, the user is informed of the lack of resources, and the system refuses to start the job.

The jobs are modeled as directed graphs. Nodes represent computation tasks, and edges represent FIFO communication channels. These graphs are not data flow graphs as used in previous chapters. In fact, after our compile time scheduling techniques have been applied to a data-flow graph, a new graph has been generated: each task in this graph corresponds to a cluster of actors (quasi-)statically ordered and annotated with budget requirements such as scheduler settings (for a TDM scheduler, this is processor cycles per TDM period) and memory consumption for both data and instructions.

Channels require memory space to store buffers, as well as network bandwidth. We did not discuss how to compute the bandwidth requirements of channels in Chap. 5, as we assumed that communication between actors was accounted for as remote stores by the producers into buffer spaces reserved at the side of the consumer. In this chapter, due to the introduction of the network, we can no longer make such an assumption, and therefore an explanation is required on where and how such bandwidth budgets can be derived.

Determining Bandwidth Budgets for Channels

With the introduction of the communication assists and the Æthereal network, communication now requires reservation of communication resources.

In terms of the temporal analysis model, a channel between two single-rate actors i and j through a producer-side CA, a number of network routers and a consumer-side CA can be represented by replacing the arc between (i, j) by a latency-rate server data flow models connected in series: a model for the producer-side CA, followed by a model for the network and a model for the consumer-side CA. The CAs are represented each by a latency-rate server model (depicted in Fig. 7.2a). The network is represented by a single latency-rate server component followed by an actor that models the latency caused by the synchronous transfer of packets through subsequent routers on the path, as explained in the previous section (depicted in Fig. 7.2b). The complete model of the temporal behavior of a channel (i, j) implemented over the network is shown in Fig. 7.2c.

These models are parametric. The execution times of the two actors c_L and c_R in the CA model are given respectively by $t(c_L) = (N_{CA} - B_{CA}(i, j)).P_{CA}$ where P_{CA} is the period of the CA time wheel (we assume TDM arbitration on the

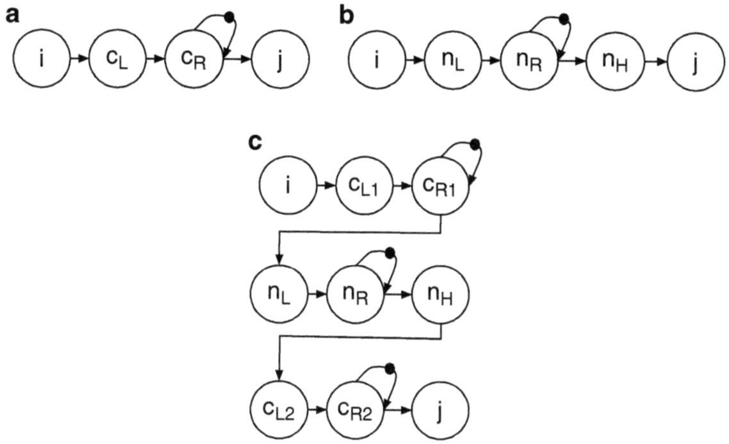

Fig. 7.2 Data flow models for (**a**) communication assist link (**b**) network link (**c**) channel from producer to consumer in different PTs

communication assist), N_{CA} is the fixed number of equally-sized slices per period that the CA scheduler provides, and $B_{CA}(i, j)$ is the number of slices allocated for link (i, j) in the CA, and $t(c_R) = \frac{z(i,j)}{B_{CA}(i,j).w_{CA}}.P_{CA}$, where $z(i, j)$ is the size in words of the tokens produced by i and consumed by j and w_{CA} is the number of words the CA can transfer per slice.

The execution times of the three actors n_L, n_R and n_H that represent the network are given by $t(n_L) = (N_{Net} - B_{Net}(i, j)).P_{Net}$, $t(n_R) = \frac{z(i,j)}{B_{Net}(i,j).w_{Net}}$, and $t(n_H) = (h(i, j) - 1)\frac{P_{Net}}{N_{Net}}$, where P_{Net} is the period the time wheel of each network router, N_{Net} is the fixed number of equally-sized slices per period that each network router provides, $B_{Net}(i, j)$ is the number of slices per period allocated to link (i, j) in a network router – it must be the same for all network routers that the connection provided to channel (i, j) traverses, and $h(i, j)$ is the number of network routers that each packet sent through channel (i, j) must traverse.

If we assume a worst case for $h(i, j) = h$, where h is the number of routers in the network, i.e. we assume that each channel may traverse all routers once, and we explicitly disallow a channel from traversing the same router twice, then all parameters in the model are application independent, except for the $B_{CA}(i, j)$ and $B_{Net}(i, j)$ parameters, which represent the CA and network bandwidth budgets of channel (i, j). Since in an SRDF graph each producer produces at most once per period, and assuming that $\mu_D \geq P_{Net}$, then can simply set the required bandwidth to $z(i, j)$ words per μ_D, which means that $\frac{B_{Net}(i,j)}{P_{Net}} \geq \frac{z(i,j)}{\mu_D}$ which implies that $B_{Net}(i, j) \geq \frac{z(i,j).P_{Net}}{\mu_D}$. Assuming that a low latency is not necessary, we can set budget $B_{Net}(i, j)$ to be equal to this lower bound (same formulas and conclusion can be drawn for $B_{CA}(i, j)$ budgets). If temporal analysis indicates that assuming values thus obtained creates an unfeasibility of throughput or latency requirements, then the budgets can be increased. Channel bandwidth budgets "withdraw" from

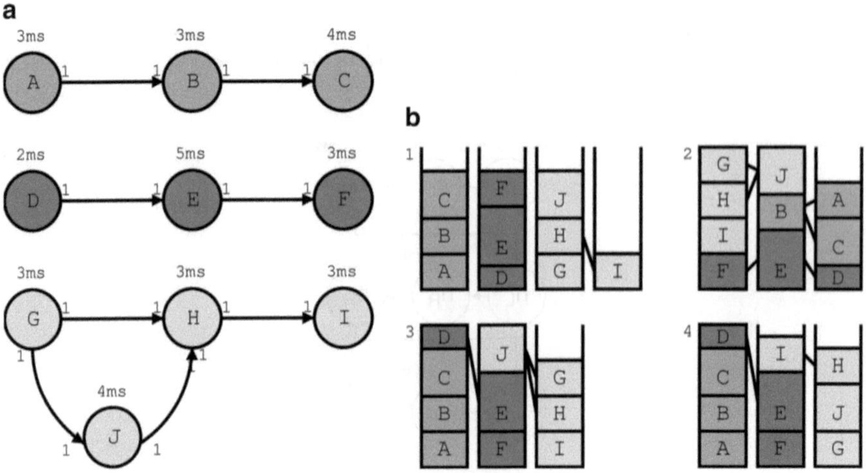

Fig. 7.3 Three job graphs with four different allocations

the same deadline extension pools as the slicing optimization phase of the compile time scheduler, and the same methods as were described in Chap. 5 can be used to determine instead or in combination with TDM slices for processor schedulers.

7.2 Resource Allocation

If a request to start a job arrives, sufficient resources need to be found and allocated for its tasks and channels. The tasks are to be allocated to PTs, and routes need to be allocated in the network for the channels between tasks on different PTs.

The problem is to find such an allocation, while trying to keep as many free resources as possible for jobs that may be started later. Also, it is a concern that computing resources may start fragmenting across the platform after many job starts and stops.

We first give a simplified example to highlight some of the challenges involved in solving the allocation problem. Then, we will describe how we formally model resource providing and resource requesting entities. We then formally define the allocation problem, discuss its complexity and strategies to solve it.

7.2.1 Motivating Example

The three jobs depicted in Fig. 7.3a are to be placed simultaneously onto three empty tiles. We assume a single-router network, and can therefore ignore for now the problems related with network topology. To illustrate the trade-offs involved, we depict several different allocations in Fig. 7.3b. In this example, for the sake of simplicity, each task in a job requires only computational resources and each tile provides only computational resources. Placement strategy 1 tries to map the whole job onto a new processor to minimize communication costs. This strategy fails to find a feasible allocation by requiring four processors to be used instead of three. Placement strategy 2 packs the tasks as tightly as possible and concentrates all the free space in a single processor. Jobs are now scattered over the processors, leading to high communication costs between tasks. If we rearrange the tasks, as shown in strategy 3, we lower the communication overhead, while keeping the same distribution of free space. Packing strategy 4 shows another possible free space vs. communication trade-off. It reduces the amount of inter-processor communication at the expense of fragmenting the remaining computing resources.

7.2.2 PT Resource Provision

A PT provides computation, memory and communication resources. This can be represented as a resource provision vector. If p is a PT, then the resources that p offers to the components of a job graph can be represented by means of a vector $r(p) := [D(p), M(p), N(p), S(p), I(p), O(p)]^T$ where $D(p)$ is the number of CPU cycles per TDM period of the processor; $M(p)$ is the amount of memory the PT offers, $N(p)$ is the number of NI queues available, from which one needs to be allocated for each FIFO channel entering or leaving this PT, $S(p)$ is the amount of CA bandwidth needed for each network connection, $I(p)$ and $O(p)$ are the upper bounds on the amount of, respectively, incoming and outgoing bandwidth for this PT.

7.2.3 Job Resource Requirements

Tasks and channels are resource consuming entities. A task requires CPU cycles as well as memory space, for both state and temporary variables (which we consider together). For a task graph $J = (V, E)$, where V is the set of tasks and E is the set of channels, this can be modeled as a vector $s(i) := [T(i), M(i), 0, 0, 0, 0]^T$ where $T(i)$ and $M(i)$ are, respectively, the CPU cycles per TDM period and the amount of memory that the task needs. All other requirements are set to 0, as they are related with communication, which will be modeled by channel requirements.

Channel requirements depend on the mapping of its source and sink tasks. If both tasks are mapped to the same PT, the channel is implemented by a FIFO buffer in the memory of that PT and we say that the channel is internally mapped. On the other hand, if source and sink are mapped to different PTs, bandwidth in the NoC has to be allocated, as well as memory at both extremities, outgoing/incoming bandwidth, network interfaces and CA bandwidth. We say that such a channel is externally mapped. The vector that represents the resource consumption of an internally mapped channel is: $i(e) := [0, M(e), 0, 0, 0, 0]^T$, where $M(e)$ is the amount of memory required to store the FIFO relative to channel e. If the edge is externally mapped, resources are required from both the source tile and the sink tile. This can be expressed by two vectors, $p(e)$ and $c(e)$, representing respectively the resource usage at source and sink PTs $p(e) := [0, M_p(e), 1, C_p(e), 0, b(e)]^T$, $c(e) = [0, M_c(e), 1, C_c(e), b(e), 0]^T$, where $b(e)$ is the bandwidth required by the channel, $M_p(e)$ and $M_c(e)$ are, respectively, the memory required to store the source and sink endpoints, and $C_p(e)$ and $C_c(e)$ represent the CA bandwidth required.

These four resource requirement vectors can be reduced to two by the following transformation. Let $w(i)$ be the amount of resources needed to host task i, i.e.,

$$w(i) := s(i) + \sum_{e=(i,b)\in E} p(e) + \sum_{e=(b,i)\in E} c(e).$$

The resources that are saved when both endpoints of a channel are mapped to the same PT is modeled by

$$\delta(e) := p(e) + c(e) - i(e).$$

Then, a set of tasks A can be allocated on PT p iff:

$$\sum_{a\in A} w(a) - \sum_{\substack{e=(a,b)\in E \\ a\in A, b\in A}} \delta(e) \leq r(p). \tag{7.1}$$

7.2.4 *Network Resource Provision*

The route through which the packets of a FIFO channel flow through the network should be allocated, and is fixed for the lifetime of the channel it supports. Such a route through the network consists of a path through the routers used and slot allocations for the slot reservation tables of each one of the routers.

In [39], path finding and slot reservation in NoC resource allocation are decoupled: first a path is found in a directed graph model of the network where nodes are routers and edges are links between routers, and given such a path, an attempt is made to do the slot allocation. If the slot allocation fails, then a new path is searched for.

We will present an approach where we use a network graph model that includes the time slots. Each router is represented by T nodes where T is the number of slots in the time wheel of the router. Each node therefore represents a specific time slot of a given router, and it has incoming and outgoing edges to every network interface and router to which its router is connected. If there is an edge is between two routers, it connects slot vertex t of the source to slot vertex $(t+1) \mod T$

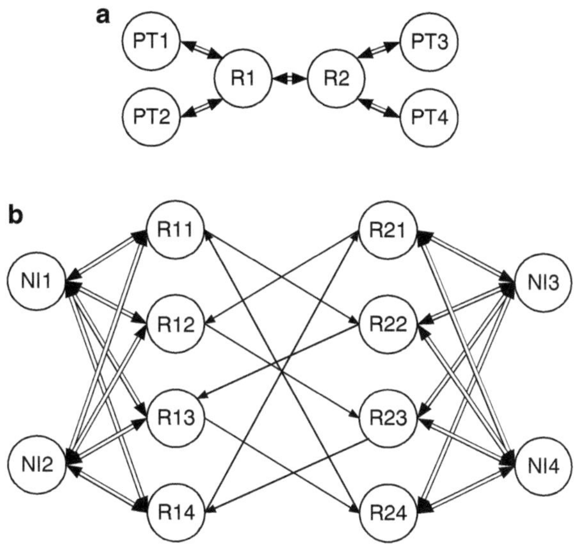

Fig. 7.4 Example of (**a**) a network with two routers and (**b**) its network graph model. A bi-directional, double-stroked arrow represents one edge in each direction

of the sink, which reflects the 1-slot delay introduced by each router. If the edge is between a router and a tile, each edge connects slot vertex t of the router to the single vertex representing the tile.

In Fig. 7.4a we depict a network consisting of two routers and four network interfaces, with two network interfaces connected to each router (we use bi-directional, double-stroked arrows to indicate that links exist in both directions). In Fig. 7.4b we depict our network graph model for this network. Again, bi-direction, double-stroked arrows indicate the existence of two links, one in each direction.

When the source and sink of a channel are mapped to different tiles, the channel requires a route through the network to be allocated between these two tiles. In the network graph model, this corresponds to finding a set of paths between the two NIs of those tiles, which together provide enough bandwidth to support the channel. In this way, finding a path through the network and a slot allocation become a single problem. The trade-off is, of course, that if the original network had R routers and L links, the corresponding graph in the traditional model has R nodes and L edges, our network graph model has $T \times R$ nodes and $T \times L$ edges.

For the whole job, more than one route will typically be allocated through the network. Because in every time slice a router can only route to each output link a single input link, these paths are not allowed to collide, i.e. to share edges. Finding several paths in the network graph model that do not collide amounts to solving the Directed Edge-Disjoint Paths Problem, which is NP-complete even in many restricted cases [52]. Instead of solving it directly by considering all paths at once, we can approximate a solution by finding routes one at a time using Shortest Path. In theory, this algorithm is capable of finding at least a fraction of $\Omega(1/\sqrt{|E_0|})$ of the routes possible, where $|E_0|$ is the number of channels allocated in an optimal solution [50]. In practice however, it usually performs much better.

7.3 Mapping Jobs

The problem of mapping a single job to the target platform can be formulated in this way:

Problem 1 (Single Job Resource Allocation). *Given a job digraph* $G = (V, E)$, *and a network topology digraph* $T = (P \cup R, N)$ *of processing tiles* P, *router nodes* R *and network links* N. *Also given a weight* $b(e) > 0$ *for each edge in* E, *representing the amount of bandwidth capacity required by channel* e. *The topology* T *is restricted by the fact that each vertex in* P *is only connected to exactly one router in* R *(assuming* $|P| > 1$*). Besides that,* P *has a valuation* $r : P \rightarrow \mathbb{N}^6$, *the resource provision of* p, *and all* n *elements of* V *have valuation* $w : V \rightarrow \mathbb{N}^6$, *the resource usage of* n, *and all* e *elements of* E *have valuation* $\delta : E \rightarrow \mathbb{N}^6$, *the resource savings for internal mapping of* e. *Does there exist an injective mapping* m *of tasks* V *to tiles* P *such that there exist edge-disjoint paths through the network (through* R *and* N *nodes) to accommodate the edges between the tasks in the mapping, knowing that each edge* e *requires* $b(e)$ *paths from the tile* $m(src(e))$ *to the tile* $m(snk(e))$, *where* $src(e)$ *is the source task of* e *and* $snk(e)$ *is the destination task of* e, *while guaranteeing that for all tiles* p *it holds that*

$$\sum_{n \in A(p)} w(n) - \sum_{\substack{e=(n,v) \in E \\ n \in A(p), v \in A(p)}} \delta(e) \leq r(p), \qquad (7.2)$$

where $A(p) = n \in V : m(n) = p$, *is the set of tasks mapped to tile* p*?*

This problem is NP-complete. We will proceed to show this in two steps: (1) show that the problem is in NP, that is, a solution can be verified in polynomial time, and (2) show that all the instances of a known NP-hard problem are an infinite subset of the set of instances of our problem. (1) Checking whether a given mapping solves Problem 1 can be done in polynomial time: first, per PT, calculate the total unmodified resource consumption by summing the resource requirements $w(n)$ of all nodes mapped to that PT; then, for each edge, check if its endpoints are in the same PT, and, if so, adjust resource consumption in that PT according to $\delta(e)$; for each PT P, check if the total resource consumption is less than $r(p)$. This has a complexity of $O(N^2)$. Check for each externally mapped edge whether the route exists. This has a maximum complexity $O(N)$ for a given fixed number of routers and slices per router in the network. (2) Any instance of off-line Vector Bin-Packing (VBP) can be reduced to an instance of our problem where items become nodes, bins become PTs and the edge set is empty. In all instances of Problem 1 where there are no edges, the sum of $\delta(e)$ in the problem formulation becomes zero in all dimensions, and the formulation of Problem 1 becomes equivalent to the formulation of VBP. Therefore a positive of VBP will be a positive of Problem 1 and vice-versa. Problem 1 is thus NP-complete.

We must also take into consideration that the solution cannot be algorithmically complex, because the resource manager must run on-line. The problem resembles the Vector Bin-Packing (VBP) problem by viewing the PTs as bins and the tasks as items. The resources provided by the PTs and the resources required by the tasks become the bin capacities and item sizes, respectively. However, there are non-trivial differences between our problem and VBP. In VBP, the items to be packed have a constant size, while in our resource allocation problem, the resources required by the endpoints of a channel depend on whether or not they are mapped to the same PT. This can be disregarded, but that would over-dimension the problem. Also, VBP does not take bandwidth usage into account, nor the available routes in the network.

Our experiments in Sect. 7.4.1 will show that, for a 1-router system where topology is not an issue, the low-complexity First-Fit (FF) and First-Fit Decreasing (FFD) algorithms give a good performance regarding the number of PTs needed and the bandwidth usage can be optimized by clustering the most heavily communicating tasks.

We then upgrade these results so that a network can be included, in a two-steps approach. An FF-based algorithm is used to map the tasks onto virtual tiles (VTs), which are assumed to be connected through a bus. Virtual tiles provide the same amount of resources as a PT would. Then, each of these VTs is mapped to a real PT, and routes through the network are allocated using shortest path.

We will describe some techniques that we combine into a solution. First, we discuss the different techniques employed for clustering tasks. Next, we argue why it is advantageous to shuffle the input of the FF algorithm. In Sect. 7.3.3, the different placement algorithms for the virtual tiles are discussed. Finally, we show how to map the VTs onto the PTs in Sect. 7.3.4.

7.3.1 Clustering Strategies

Clustering can be applied before and during packing. If applied before, a set of tasks is replaced by a single larger task, thus forcing the set to be placed on the same PT. We will use a greedy heuristic, which orders the channels in order of decreasing bandwidth requirements, and replaces each pair of endpoints by a single large task if the resulting task fits on an empty PT. This is repeated until a predefined percentage of the number of channels in the job is contracted, or until no more channels can be contracted. This can lead to an increased number of required PTs if the clustering is done too aggressively.

If clustering is applied during packing, it works together with the packing heuristic: when the packing heuristic packs a task to a PT, it will then try to pack adjacent tasks to the same PT before trying to pack other tasks. This modified First-Fit packing algorithm will be called First-Fit with Clustering (FFC). Our experiments Sect. 7.4.1 show that FFC performs significantly better than other FF variants for virtual tile placement.

7.3.2 *Shuffled Input*

Both FFC and the pre-clustering strategy behave in a deterministic manner and without backtracking. This gives these algorithms a single shot at finding an acceptable solution. To incorporate backtracking is non-trivial. The problem it has to solve is NP-complete, so a solution which uses full backtracking can take an exponential amount of time. Instead of backtracking, we use randomization to generate multiple distinct solution candidates. The First-Fit algorithm considers and places the tasks one at a time, which allows a randomization of the input to yield different mappings. An unsuccessful mapping of the tasks onto a set of bins using one ordering of the tasks can thus occasionally be corrected by considering a different ordering of the tasks. If the First-Fit algorithm clusters during packing, it still just starts to map the 'next' task when it has filled a PT, which keeps it sensitive to the original ordering of the items.

7.3.3 *Virtual Tile Placement*

The First-Fit packing algorithm packs tasks into the PTs, but considers empty PTs to be equal and uses them in the order they are presented. However, due to the presence of the network and the bottlenecks that can be formed between two PTs, it is not efficient to use just any subset of the empty PTs. It is favorable to map both endpoints of a channel as close to each other as possible. If they do not fit on the same PT, they should preferably be placed on PTs close to each other, for instance on two PTs connected to the same router.

We want to accomplish this by employing a two-step system using virtual tiles (VTs). The packing algorithm maps tasks to a VT, as if the VTs are interconnected through a bus. Then, each VT is assigned to a real PT by in the second step, and routes through the network are allocated. We compare three methods to do this assignment:

On-line placement: As soon as the packing algorithm needs a new, empty tile, a PT is selected. Because it is unknown which tasks are going to be assigned to this new PT, there is little information for which PT to select. As a heuristic, to minimize the usage of network resources, the PT closest to those already used is chosen, using the sum of distances to the already used PTs as the distance metric.

Semi on-line placement: The packing algorithm is capable of filling the VTs one by one. Once filled, the content of a VT does not change, so the VTs can be mapped to PTs immediately. This allows a full VT to be placed near those PTs which contain tasks connected to those in it. The number of links required in the network to allocate the channels to that VT is used as the metric for distance.

Off-line placement: All tasks are packed into a set of VTs first. This set of VTs is then mapped to the PTs, minimizing the number of links used.

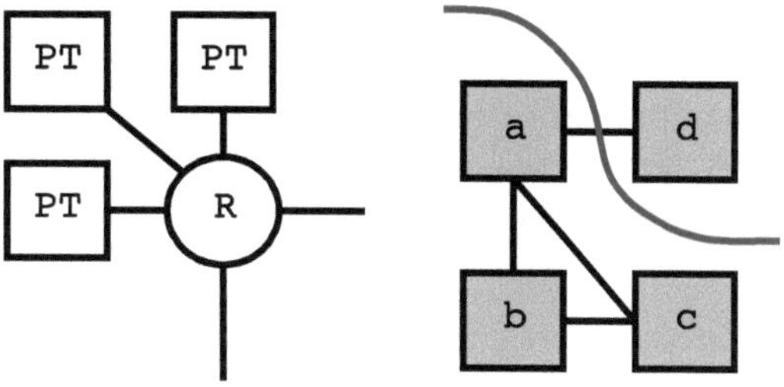

Fig. 7.5 A router connected to three PTs (*left*) and four VTs connected by channels (*right*)

The described placement methods assume an empty system. If the system is not empty, it is preferable to fill up partially filled PTs first to avoid fragmentation of free space. To adjust all algorithms to be able to cope with a non-empty system, the placement methods let the packing algorithm fill up the partially used tiles first. Thus, the location and space limitations for the VTs representing these locations are known beforehand.

7.3.4 Bisection According to Kernighan-Lin

The off-line placement algorithm maps the VTs to PTs such that the number of links used is minimized. As will be shown, this problem contains NP-complete subproblems, so a heuristic is needed. For our purposes, the network consists typically of 4–12 routers, with each of which is attached to several PTs. The heuristic we use tries to map the VTs to PTs such that most of the required links will be mapped over a single router. To accomplish this, the placement algorithm selects a router and counts the number of empty PTs around it. The algorithm then selects the same number of VTs, minimizing on the bandwidth required between the selected and not selected VTs. These selected VTs are then mapped to the empty PTs. Once such a cut is found and the selected VTs are mapped to PTs, a new router is selected. For example, take Fig. 7.5, which shows a router surrounded by empty PTs as well as a set of unmapped VTs to map. Because the router is connected to three PTs, three VTs are cut off. A cut is made such that the number of links to the rest is minimized. These VTs (a, b and c) are then mapped on the PTs around R in any order. This process is repeated until the placement algorithm runs out of VTs (or empty PTs, in which case the algorithm fails to find a valid placement). The order in which the routers are selected is predefined; we acknowledge this as a point of future research.

The off-line placement algorithm is thus required to find a subset of VTs of fixed size, with a minimal cut to the rest of the VTs. This problem contains the Minimal Bisection problem, which asks to divide a graph into two equal pieces with a minimal cut and which is NP-complete [20]. In our case, two partitions of unequal size are needed, say of sizes a and $|V| - a > a$. To accommodate this, we add a clique of size $|V| - 2a$ to the graph. The vertices of this clique are not connected to any vertex outside the clique, and the edges of the clique have an infinite weight.

When Minimal Bisection is applied, this clique will end up in one of the partitions, together with other vertices. Any other solution yields a cut of infinite weight. The rest of the vertices will thus form a $a : |V| - a$ split.

A well-known heuristic for Minimal Bisection is the algorithm designed by Kernighan and Lin (KL) [49]. It starts with any bisection and iteratively improves on it as follows:

1. Create any bisection of V into P_1 and P_2 with $|P_1| = |P_2|$.
2. Set all vertices to 'unlocked'.
3. Let S be a list of pairs of tasks, and let G be a list of integers. Set S and G equal to the empty list.
4. For every pair of unlocked vertices $a \in P_1, b \in P_2$, calculate D_{ab}, which is the reduction of the weights of the edges in the cut, were a and b to be swapped.
5. Find a pair $a \in P_1, b \in P_2$, for which D_{ab} is maximal. Append (a, b) to S. Append D_{ab} to G. Swap and lock a and b. The value $\sum_{i=0}^{|S|-1} G_i$ now denotes the decrease in cutsize so far.
6. Repeat the previous two steps until all vertices are locked.
7. Find $n := \arg\max_{0 \leq n \neq |S|} \sum_{i=0}^{n-1} G_i$, which represents the moment at which the cutsize was the lowest.
8. Swap back all pairs in S after index n.
9. Repeat steps 2–8 while there is an improvement $\left(\text{i.e. while } \sum_{i=0}^{n-1} G_i > 0\right)$.

Further modification of the KL algorithm is needed to account for the fact that we cannot swap VTs that are already mapped to PTs. To ignore these tiles means to ignore their channels to VTs that still need to be mapped. Rather, they can be taken into account by locking the already mapped VTs at step 2.

7.4 Experiments and Results

In this section we will present a set of experiments to evaluate the performance of several combinations of the presented algorithms.

First, we will present results in the simplest case, where there is only one router in the system, and therefore topology has not to be taken into account, and there the virtual tile mapping corresponds to the physical tile mapping. This case is relevant, because a single router is capable of handling up to eight tiles, which is more than enough for the size of current SDR systems.

After that, we will consider the case where more than one router is necessary. We will start by comparing the performance of the simplest variants on several network topologies. Then, the difference in performance is measured between the on-line, semi on-line and off-line VT placement algorithms. After that, we will evaluate the gain in performance when the ordering of the input tasks is shuffled in order to obtain several solution candidates. Finally, the best of these alternative algorithms is tested on a system with a high load.

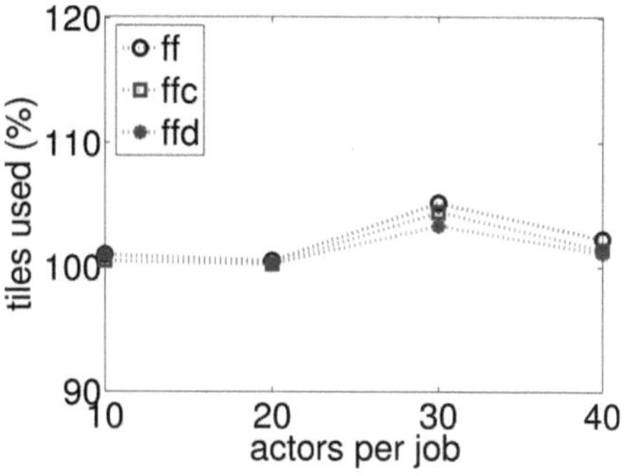

Fig. 7.6 Percentage of tile utilization with respect to the total amount of resources required as a function of the number of tasks per job

As a job set, we used synthetically generated graphs. These graphs have at most 100 tasks, and contain a number of channels similar to the number of tasks. We divided them in four test sets depending on task and channel bandwidth requirements. Because of lack of space, we only show the results for graphs containing tasks with resource requirements in the range of about 4–14% of the resources available by PT and edges require between 2 and 7% of PT resources and bandwidth between routers. For smaller tasks and edges, the results are better. The complete results can be found in [70].

All tests were performed against varying rates of clustering, from contracting 0 to 60% of the channels. Trying to cluster more than 60% of the channels failed in all cases.

7.4.1 Systems with One Router

VBP Algorithm Test

We first tested mappings of single jobs on an empty system for several low-complexity VBP algorithms: First-fit (FF), First-Fit Decreasing (FFD), Best-fit (BF) and Best-fit decreasing. To this we added our algorithm that performs clustering while packing. It is a First-fit algorithm that assigns a task to a PT, and then looks at the neighbors of that task and tries to put them in the same PT. We refer to it as First-fit with Clustering (FFC).

The results were compared with an exact solution for VBP (which disregards $\delta(e)$ resource savings) obtained by a Branch-and-Bound (BB) algorithm. The result is shown in Fig. 7.6. The BF variants are not shown for simplicity sake – the results were similar to the FF variants.

All algorithms performed similarly well in terms of PT usage, only using at most about 5% more PTs than the BB algorithm. For this reason, we decided that there was no need to experiment with more complex VBP algorithms. We also discarded the BF variants, which are more complex than the FF variants. FFC still got better

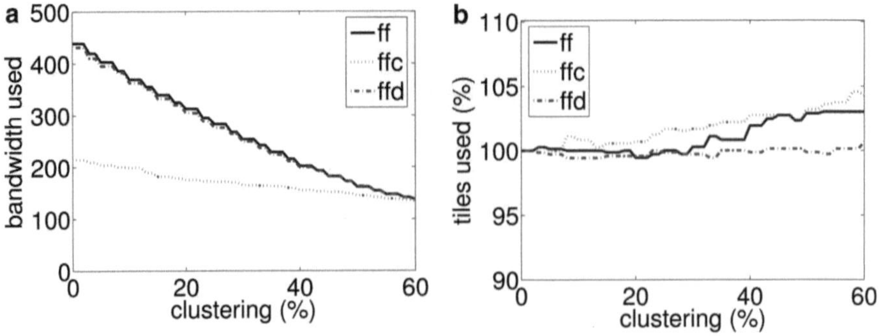

Fig. 7.7 Effects of pre-clustering on bandwidth and tile usage

results than other algorithms as it allows for more resource savings. It turned out that our results were much better than the theoretical worst-case performance [14] of FF and FFD of about 553% increase on the number of bins used for 6 dimensions.

Bandwidth and Clustering

We introduced clustering before packing to try to improve the bandwidth usage of FF and FFD. We used a greedy clustering algorithm that orders all edges non-decreasingly by their bandwidth requirement. For each edge, it clusters the two endpoint tasks if the resulting cluster still fits in one PT. Only a certain percentage of edges is allowed to be contracted into a single cluster, allowing a trade-off between saved bandwidth and granularity of the clusters. The Clustering Before Packing (CBP) algorithm attempts to contract 0–60% of the channels, in 1% increments. Sometimes it is not possible to cluster the requested percentage, as clusters become too big to fit in a PT. In that case, as many nodes as possible are clustered. Trying to cluster more than 60% of the channels never succeeded.

Figure 7.7a, b show the results of these experiments. In Fig. 7.7a we can see how much bandwidth gain we can achieve by using CBP. FFC saves about 44% of bandwidth when CBP is not used, when compared with FF and FFD. CBP also enhances the bandwidth saves that FFC already provides. At the highest percentage of CBP, the bandwidth usage of FFC and the other two algorithms becomes indistinguishable. The reason for this is twofold. First, with the increase of the average size of clusters it becomes more difficult for FFC to combine neighboring tasks on the same PT. Second, the greedy algorithm of CBP can outperform FFC since it uses global knowledge of the job, while FFC only tries to add neighbors of tasks already mapped.

The gains in bandwidth usage provided by heavy clustering are counterbalanced by a slight increase in PT usage, as depicted in Fig. 7.7a. This is still less than a 5% increase even for maximum (60%) clustering.

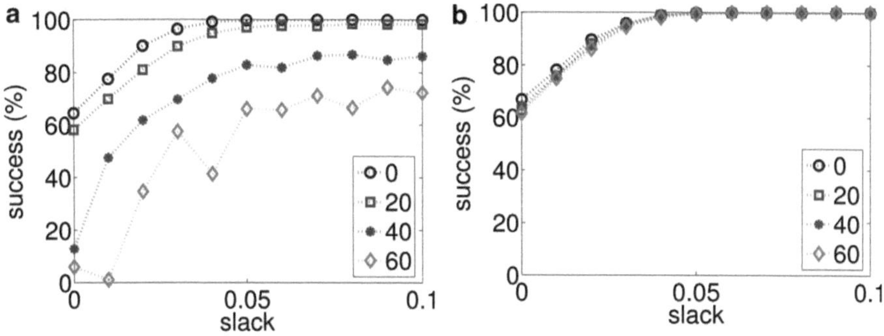

Fig. 7.8 Mapping successes versus slack (**a**) without fall-back and (**b**) with fall-back

Stress Test

The previous tests assumed an empty system and tried to reduce the resource usage. We also need to know how the system fares under a heavy load, i.e., if the PTs have barely the amount of resources needed by the job that presently tries to enter the system, what is the likelihood of a mapping to be found. Also, we need to know how scattered allocation will affect the capability of the system to use its resources. To answer these questions we designed a stress test that constantly forces the allocator to map jobs under heavy load. This test works as follows.

First, we choose a job set J. For each job $j \in J$, the sum r_j of the processor cycles and memory requirements of its tasks and channels is calculated. Let t be the sum of all processor cycles and memory space still available in the system. Vector t is updated whenever a job is started or stopped. The system starts without any jobs running. A job j which is not already running, is selected at random, and is requested to start iff $(1 + \sigma)r_j \leq t$, where $\sigma > 0$ is an amount of slack. The slack parameter σ is introduced as it is very unlikely that any bin-packing algorithm (even an optimal one) is capable of filling up the PTs to 100%. If the inequality $(1+\sigma)r_j \leq t$ does not hold, running jobs are stopped at random until the inequality does hold. The resource allocator is then requested to start the job.

The resource allocator receives this job start request and tries to map the job onto the unused resources. Its success or failure to do so is noted, and the next job is randomly selected. This is repeated for 10,000 iterations for several values of slack σ. During the first steps of the stress test the selected jobs will gradually fill up the system. After a while, the resource allocator will be trying to map new jobs onto a heavily loaded system. We can then measure the number of successful mappings.

We have chosen several test job-sets. For each of these, we ran the stress test for 10,000 iterations. The results shown are for the FFC algorithm, which provided the best results in the previous tests. The plot in Fig. 7.8a shows, for different percentages of CBP applied, the percentage of successes of the FFC algorithm plotted for different amounts of allowed slack. There are several things worth noting

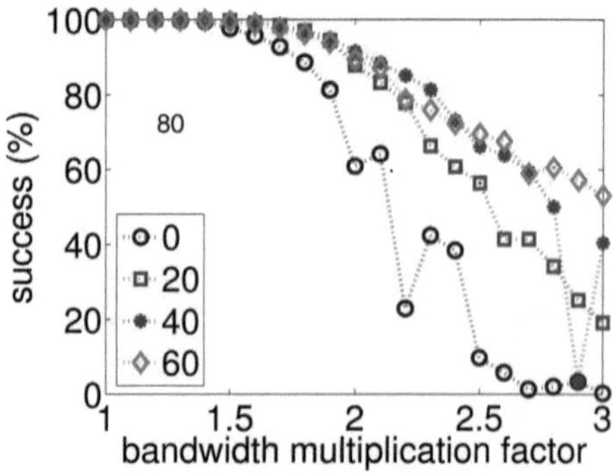

Fig. 7.9 Mapping successes with increased bandwidth requirements

here. The first is that, if no CBP is used, less than 10% of slack is enough to virtually eliminate all mapping failures. The second is that CBP does not seem to work very well. A case analysis showed that the reason for this is that CBP creates clusters that cannot be combined well on half-filled tiles, thus leaving large gaps which the similarly large clusters of the new jobs cannot fill. In the previous experiments, with an empty system, this problem could not be seen. A third observation is that the graph occasionally spikes downwards with the increase of slack. Case analysis showed this is due to the stress test getting stuck in a local minimum, and a row of jobs fails to be mapped. This effect is magnified at high CBP percentages, where fragmented mappings pose more of a problem.

We decided to try to fix these problems by resorting to a fall-back mechanism: if the RA cannot map the clustered version of the job, it tries to map the original, unclustered version. The results can be seen in Fig. 7.8b. With this modification, CBP does not affect negatively the performance anymore and can even improve it slightly. Clustering optimizes bandwidth usage, not PT usage. To see if CBP is useful, the bandwidth requirements of all jobs were scaled up and a slack of 10% was allowed. The results are shown in Fig. 7.9. CBP does help in decreasing bus usage. With a factor of 3 increase in the bandwidth requirements, a clustering of 60% still allows for a success rate of 60%. Without any clustering, the success rate has gone down to almost 0%.

Discussion of Results

For a 1-router system, the FF family of VBP algorithms provided a simple yet effective base for an allocation heuristic. The best solution is a two-step approach: first, apply CBP and try FFC; second, if this fails, apply FFC with the original graph. This combined solution is capable of utilizing at least 95% of the available PT resources, if the bus bandwidth does not become a very scarce resource.

Table 7.1 The topologies considered

Topology	Size	Links/router	Tiles	Routers	Links
Ring-3	3	10	24	3	54
Ring-4	4	8	24	4	56
Ring-6	6	6	24	6	60
Ring-12	12	4	24	12	72
Mesh-6	2 by 3	6	22	6	58

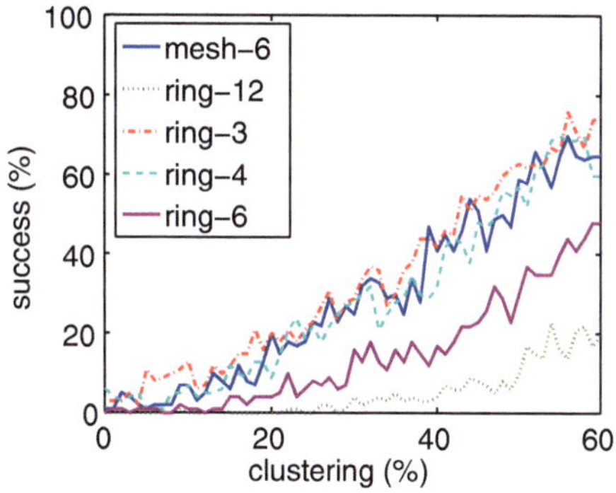

Fig. 7.10 Successes per topology

7.4.2 Systems with More Than One Router

For systems with more than one router, we tested several topologies, shown in Table 7.1. These topologies will be compared first. In a ring topology, routers are connected to neighboring routers in a one-dimensional grid. In a mesh topology, routers are connected to north, south, east, west neighboring tasks in a two-dimensional grid.

Effects of Topology

For each of the topologies of Table 7.1, the FFC algorithm was chosen to try to map an instance of 100 tasks onto an empty system. We used the on-line virtual placement method. This was tried for 100 instances, each mapped onto an empty system. The percentage of jobs that could successfully be mapped is shown against the clustering percentage in Fig. 7.10.

We observe several facts from this figure. First, clustering is necessary to be able to map any of the tested jobs onto any of the tested topologies. Secondly, the larger rings (ring-6 and ring-12) give bad performance (50% success rate or lower), while the performance for the other topologies is similar (up to 70% success rate). This could be due to the fact that in ring-3, ring-4 and mesh-6, the average distance between two PTs is low. Hence, less congestion is likely to occur, which would hinder a successful mapping. What would be the optimal topology heavily depends on the cost function. In our case, routers contain an all-to-all interconnection grid,

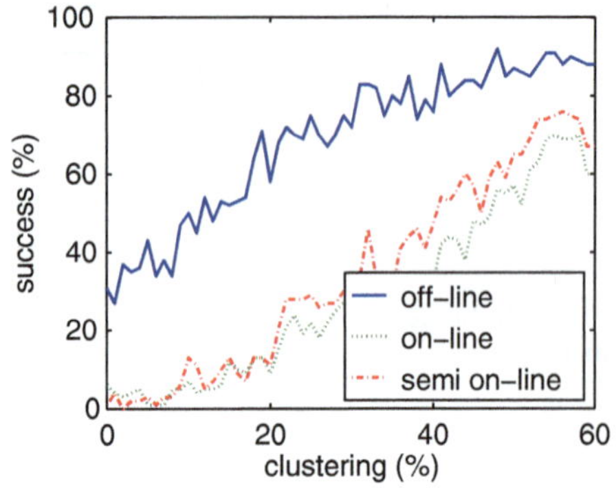

Fig. 7.11 The percentage of successful mappings as a function of the clustering percentage for different placement algorithms

thus having a cost quadratic in the number of links they are connected to. Each link and router also bears costs, further complicating judgment. For the rest of this chapter, we chose the ring-4 topology as a trade-off between number of routers and the number of links of each router, balancing the costs.

The rest of the tests will focus on the ring-4 topology.

Virtual Tile Placement

The best performance observed in all topologies was only a 70% success rate for finding a mapping for a large job on an empty system. This percentage increases if a different form of VT placement is used. In Fig. 7.11, the success rate is shown for the same job set on a ring-4 topology, using different VT placement methods.

The difference between the on-line and semi on-line placement methods is marginal. This can be explained by the little knowledge available and used by both algorithms when asked to place a VT. The on-line algorithm has no knowledge of the contents of the tile, so has little information to perform optimizations, other than choosing a location close to the tiles already mapped. The semi on-line algorithm knows which tasks are mapped to the VT it has to place, but does not know the location of all the tasks which have channels to it. As can be seen in Fig. 7.11, this extra knowledge does not increase the performance significantly.

The off-line VT placement algorithm does prove to be a substantial improvement over the on-line one. The success percentage is raised across all clustering percentages, reaching 90%. Even though the performance is better, clustering is still required to reach the higher chances of success.

The difference in placement algorithms is not only visible in the percentage of successful mappings, but also in the number of links used by successfully mapped jobs. The percentage of links used in the network, relative to the number of links the basic on-line placement algorithm uses, is shown in Fig. 7.12. The spikes at low clustering percentages occur because there are relatively few successful mappings over which the average is taken. Again, the on-line and semi on-line algorithm differ little in link usage. The off-line algorithm attains a profit of 15% in the number

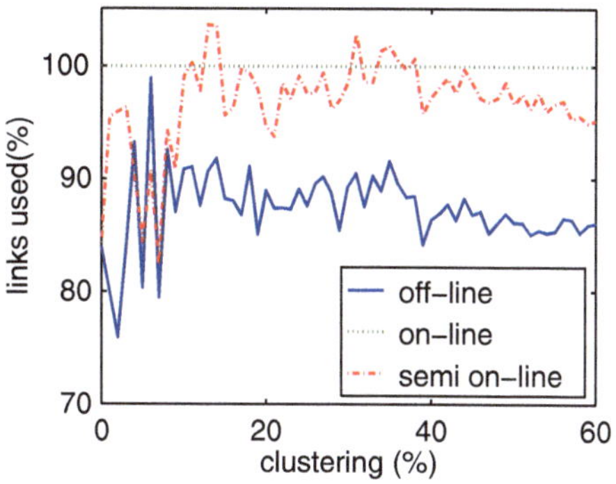

Fig. 7.12 The percentage of links used as a function of the clustering percentage, relative to the on-line algorithm

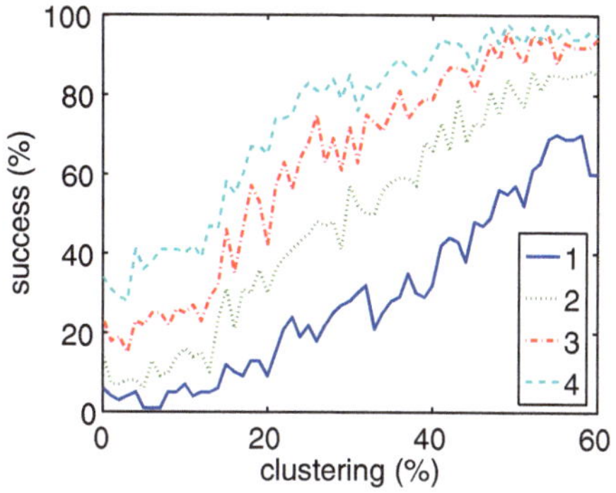

Fig. 7.13 The percentage of successful mappings as a function of the clustering percentage. Each line represents a certain number of randomized attempts

of links used, which is decent considering many routes that are optimized will be reduced from length 3 (PT–router–router–PT) to length 2 (PT–router–PT), for which only a gain of 33% in length is possible. This directly implies profit in terms of power consumption and amount of free resources left for other jobs.

Using Shuffled Input

Another way to increase the chances of success is by trying to map several random permutations on the ordering of the tasks to map. This results in a favorable increase as is shown in Fig. 7.13. The figure shows the percentage of successes when 1–4 random permutations are tried, increasing the success rate up to over 95% when 4 permutations are tried.

Stress Test

The previous tests assumed an initially empty system. However, our solution should behave well under heavy load, i.e., if there are barely enough resources available in the system when there is a request to start a job.

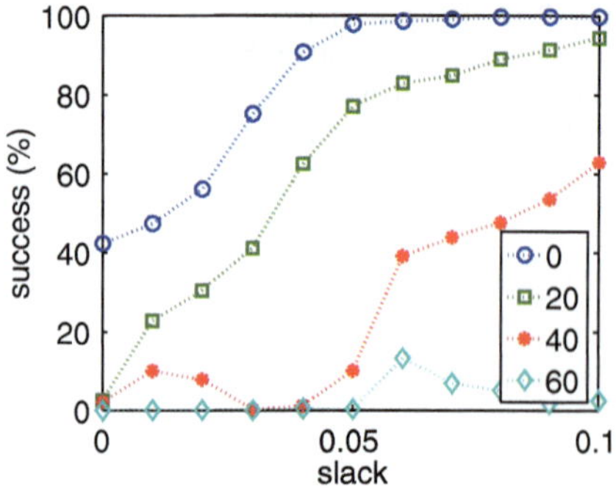

Fig. 7.14 The percentage of successful mappings as a function of the slack σ. Each line represents a certain clustering percentage

We tested the performance under heavy load by taking the ring-4 topology and performing the stress test, as for the 1-router case.

The result of this test is shown for several clustering percentages in Fig. 7.14. At $\sigma = 0$, there is no slack, implying the resource allocator has to be capable of starting a job no matter how much the available resources are scattered. This is not always possible, so we cannot expect the resource allocator to always succeed. As more slack is added, the resource allocator has more redundant free resources it can use, increasing the success rate dramatically. With only 5% slack, all requested jobs could be started if no clustering is used.

Another observation in relation to Fig. 7.14 is that, as in the single router case, clustering actually has a negative impact on performance. The reason for this is that if clustering is applied, larger tasks are created that require larger portions of free space on a PT. This requires the available resources to be concentrated on fewer tiles, something that is less likely to happen. Because clustering reduces bandwidth usage (and thus power usage) as well as increases the feasibility of mapping bigger jobs, we did not want to discard it. Instead, we use the same fall-back mechanism as in the single router case, in which first the job is clustered and an attempt is made to map it. If this fails, we re-try to map the job without pre-clustering. Any remaining attempts with shuffled input are done with the unclustered version. Again, this allows us to profit from the positive effects of clustering when possible, while being able to also profit from the higher success rate of trying to map fine-grained, unclustered versions of the jobs. The results of applying the fall-back mechanism are shown in Fig. 7.15.

Note the slack factor introduced only stretches the computing cycles and memory requirements. It is assumed bandwidth is not a scarce resource. If it is, clustering does help to reduce the bandwidth usage, but cannot prevent the allocator from starting to fail.

The results shown are for large tasks and heavy edges. The results for lighter versions of either or both can be found in [70]. All algorithms perform better when the tasks or edges are smaller, due to a finer granularity. The finer the granularity of the tasks, the easier it is to find allocations for them in the PT.

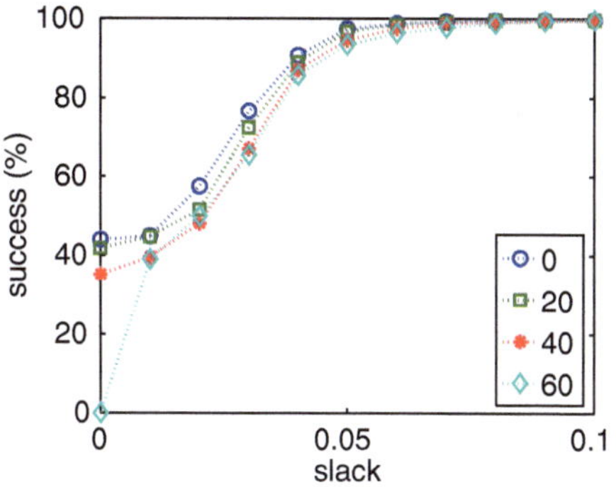

Fig. 7.15 The percentage of successful mappings as a function of the slack σ when a fall back to an unclustered version is used. Each line represents a certain clustering percentage

7.5 Further Reading

An extensive survey or the traditional techniques for scheduling and resource allocation in embedded multiprocessors can be found in [81]. It describes techniques from fully static to fully dynamic scheduling. This work does not consider the case in which tasks arrive and leave at runtime, as in our case. The same holds for techniques that compute task assignments at design-time, such as [89].

In CPA [67], it is considered that jobs may start and stop at any time. Each job mix has its own schedule, which is calculated at compile time and stored in a look-up table. This approach is not without problems. The number of potential job combinations is exponential in the number of jobs, and switching from one configuration to the other could mean a non-trivial processor migration of running tasks. Moreover, it assumes that all tasks are known at compile time.

The literature on task allocation for tasks with a periodic requirement is extensive, and covers many combinations of constraints. Previous approaches either do not consider any network [4, 29, 59, 63, 93], consider only a bus topology [5], require tasks to be migratable between processors [29], or require a solution to be computed at compile time [67, 89]. Hansson et al. [39] consider the network, but do an off-line computation of the network slot tables, provided the assignment of tasks to processors is given.

In [66], the authors propose a technique similar to the one described in this book to model the network-on-chip resources as a graph where each node corresponds to a slice in each one of the routers, and each edges corresponds to a link. This work, however, does not consider processor allocation, assuming it is given a priori.

7.6 Conclusions

In this chapter, we presented a set of heuristics to map real-time streaming jobs onto a homogeneous multiprocessor system containing a network with up to 12 Æthereal routers. We present a model that represents hardware constraints, and the resource requirements of jobs. Within this model, it is possible to reason about temporal guarantees, which allows us to map jobs onto the system such that temporal constraints are met.

We showed how to model the Network-on-Chip in such a way that time-multiplexing can be taken into account by representing each router as a number of communication nodes, one per time slice. The problem of finding a path through the network is then merged with the problem of finding a valid slot allocation table per router. The joint problem can be seen as a disjoint paths problem.

To map a job, its tasks and channels are packed onto the tiles using a First-Fit strategy, after which routes through the network are allocated. We showed it is beneficial to use a two-step approach. First, we map the tasks onto virtual tiles (VTs), and then we map the VTs onto the actual tiles. This raised the chance of finding a feasible mapping, and reduced bandwidth usage.

We also showed that chances of finding a feasible mapping can be further increased by trying several random permutations on the ordering of the input.

Our stress test shows how our resource manager performs under heavy load. One interesting remark is that although clustering is essential to map some jobs, and greatly decreases bandwidth usage, it has an adverse effect on allocation success once the system is close to full. Because of this, we devised a mechanism by which two versions of the job are tried out in the mapping: first a clustered version, and, if this fails, an unclustered version. This allowed for a substantial increase in the percentage of successes during the stress test.

When operating under heavy load, the described approach seems to be capable of allocating 95% of the resources available on the processing tiles, assuming that bandwidth does not become a scarce resource.

Chapter 8
Demonstrator

The demonstrator prototype presented in this chapter was part of a larger demonstrator built by Nokia, NXP and ST-Ericsson, where our baseband software framework is integrated with the Radio Computer functional architecture proposed in [1]. It shows how the concepts we proposed can be worked out into an execution platform, even when the hardware was not designed to meet some of the requirements of our software framework. It also shows how the custom components of our software framework can be implemented while re-using a substantial amount of existing off-the-shelf software.

As for the hardware, it uses an NXP prototype board for single-radio, integrating 2 EVPs, 3 ARMs, and hardware accelerators. All processors run at low clock frequencies: the EVPs at 183 MHz and the ARMs at 61 MHz. Each core has private data and instruction memories, and no cache or memory management unit. Connectivity is provided by a tree of AXI busses. Worst-case memory access latencies can be computed, since the AXI slave-side arbiters are Round-Robin. DMAs are not used, since bursts cause large worst-case access times through the arbiters. The Resource Manager in this implementation supports at most 5 transceiver types and 5 jobs, a number that was decided as a result of use-case analysis.

8.1 Streaming Framework: Sea-of-DSPs

In our demonstrator, Sea-of-DSPs (SoD) is a low-level multi-processor streaming OS that was developed initial at Philips and later at NXP Semiconductors. SoD was developed as a lightweight streaming framework for a multi-DSP chip.

SoD provides runtime support for the following features:

- **Streaming Kernels**: dynamic single-core schedulers on which tasks can be started and stopped; the default SoD scheduling policy is non-preemptive, round-robin;

O. Moreira and H. Corporaal, *Scheduling Real-Time Streaming Applications onto an Embedded Multiprocessor*, Embedded Systems 24, DOI 10.1007/978-3-319-01246-9_8,

- **Streaming API**: FIFO communication primitives;
- **Network Manager API**: can create, delete, resume and suspend tasks in any of the Streaming Kernels; and it can create or delete FIFOs to connect tasks.

SoD is designed to function with minimal demands on memory footprint and cycle budget.

Internal memory management is done by reserving, at compile time, configurable memory pools for each object type. This results in predictable behavior and no fragmentation. SoD has been ported to several processors: ARM, TriMedia, EVP (VD32040), EPICS7B, Tensilica Vectra LX DSP, HiFi 2. It can run in emulation mode on Windows and Linux. It has a low memory footprint: the Network Manager is less than 10 KB of ARM7 thumb code and the SoD SK (Streaming Kernel) code is less than 1 KB.

The streaming API of SoD relies on shared memory access between communicating cores. SoD allows FIFO buffers to be placed either at consumer or producer side. It also supports the usage of duplicated FIFO administration at consumer and producer side, to avoid remote read operations.

An embedded hardware platform typically has its memory scattered across the system. Each core has local data and instruction memories, and shared system memories are also available. SoD offers several choices for the location of the FIFO data and the FIFO administration. Dependent on the latency of reads and writes to each of the memories, the most optimal partitioning can be chosen when configuring SoD for a particular hardware platform. For example: in the case of a task on the ARM communicating to a task on an EVP, it is optimal to store the FIFO buffer and administration in EVP local data memory. This because of the very high latency the EVP has in writing to external/remote memories. In the case where two DSPs have to communicate, each DSP can read and write the local memories of the other DSP but remote reads will cause a much higher latency than the write cycles (posted writes). Thus the FIFO buffer is placed in the local memory of the reading task.

SoD also minimizes latency/stall cycles caused by remote reads by employing duplicated FIFO administration. If it is configured with that option, a copy of the FIFO administration table is placed at the local memory of both consumer core and producer core. Using this scheme, no remote reads are required, since consumer and producer tasks simply check the state of the FIFO by reading their local copies. Of course than upon any change to the administration of the FIFO requires both producer and consumer to update the local copy and the remote copy, but this only requires single-word remote writes, which have a considerably lower latency than remote reads.

8.2 Resource Manager Implementation

The Base-band Resource Manager (BB-RM) uses the services provided by the SoD Network Manager API. SoD already manages some resources. It allocates memory for FIFO capacities, and creates tasks and FIFOs. It does not handle jobs as groups of interconnected tasks that must be admitted or rejected atomically. It merely finds and reserves resources for a single task/channel at a time. It is not aware of timing requirements and does not perform scheduling tests. It does not chose task to processor mappings. BB-RM must take care of all these tasks.

BB-RM enforces strict admission control and resource reservation. When requested by Central Resource Manager (see diagram in Fig. 2.2) to start a transceiver, it performs admission check, trying to find a feasible processor assignment for all tasks of the transceiver.

As described in Chap. 7, the resource requirements of a transceiver are represented by a graph. Nodes represent the two types of resource consuming entities: tasks (quasi-statically ordered clusters of MCDF actors) and channels (inter-cluster MCDF FIFOs). Each node has a requirements vector. The vector fields are different from the model presented in Chap. 7, as the communication infrastructure does not contain communication assists or a network-on-chip. For a task mapped to a TDM scheduler, the vector fields are: processor type, period, slice size, code memory read-write memory, and read-only memory. For NPNBRR schedulers, slice size is replaced by worst-case execution time. Admission control checks if the sum of execution times/slices of tasks assigned to the core plus the execution time/slice of the new task is lower than the deadlines of all tasks involved. For channels, the vector fields include buffer memory on the consumer and producer side. The representation of resource requirements by a vector of scalar fields is flexible and easy to port to different hardware platforms.

Processor assignment uses the adapted first-fit vector bin-packing algorithm described in Chap. 7 to fit the resource requirements of nodes to the resource provision vectors of processors. The algorithm assumes that memory is not fragmented. After deriving a processor assignment, BB-RM invokes a memory allocation procedure per segment. If memory allocation is successful, all tasks and FIFOs are instantiated. If instantiation succeeds, the transceiver is admitted. If not, BB-RM rolls back all reservations and informs CRM of an in-feasibility. When CRM requests a job to stop, all its tasks are destroyed, and resources freed.

Since transceivers are compiled independently and processor assignment is done at admission control, BB-RM supports dynamic loading with dynamic linking to the Streaming API. The code in the transceiver configuration is compiled with relocation information that BB-RM uses to link and load the code at the addresses given by memory allocation.

BB-RM invokes a loader for each task. It provides it as input the start address and size of each allocated memory segment. The loader requests an object file from the Configuration Manager for each task. This file is parsed, and the values for symbols left unresolved at compile time are now filled in. Code and read-only data are written into the memory of the processor to which they were mapped. Symbols of newly loaded modules are added to the internal symbol table of the framework so that they can be found by modules loaded later.

8.3 Prototype Figures

In this section we evaluate the costs and overheads of our approach as evidenced by the demonstrator.

8.3.1 Memory

The data memory of BB-RM dominates the footprint, requiring 49.8 KB. Of this, 28.6 KB are spent in transceiver configuration. This requires two-fifths of the memory of an ARM (128 KB). However, there is plenty of room for optimization. We set the number of components per transceiver to 50, to be able to store complete unclustered transceivers. This was needed for testing, but oversized: for the WLAN receiver, there are only three components after clustering.

Memory allocation also consumes considerable space, at 13 KB. This is because of the fine granularity (32 bits) at which blocks of memory are allocated. The SoD Network manager consumes a total of 25.6 KB of data memory, for its own configuration information. The Streaming Kernel requires 8.6 KB, both on EVP and ARM. For instruction memory, BB-RM requires 24.5 KB, and the Streaming Kernels require 1.6 KB.

8.3.2 Performance of BB-RM

For a job containing 16 tasks and 22 FIFOs, the admission time is 201 ms, where 196 ms are taken by memory allocation. Processor allocation takes negligible time (less than 5 ms). According to the specification, a new job should take less than one second to start, so these values are acceptable. Furthermore, the clock speeds in the demonstrator are very low compared with what is currently available. If a need would arise for BB-RM to execute faster, the main area for optimization is of course the memory allocation. As already mentioned, the current granularity of memory allocation is in blocks of 1 word, using a Best-fit algorithm, that must scan all "holes" in the memory of a processor to find the best allocation. This guarantees minimal memory fragmentation, but makes the process of memory allocation very slow. Memory allocation could be made a lot faster by choosing a coarser granularity of memory block allocation.

8.3.3 Scheduling Overhead

The scheduling overhead of the Streaming Kernel per task, for the EVP is 99 ns at 183 MHz. When a task is activated, it checks for input data and output space. A check for data/space takes 135 ns, and an update of a FIFO management table

Table 8.1 Synchronization overhead per task

Standard	Schedule period (μs)	Task overhead (%)
WLAN 802.11a	4	16.3
LTE	71.4	0.32
TD-SCDMA	112.5	0.58
UMTS	133	0.49
DVB-T/SH	935	0.07

takes 141 ns. The total synchronization overhead of a task that reads one FIFO and writes one FIFO is $135 \times 2 + 141 \times 2 + 99 = 651$ ns. Table 8.1 depicts the synchronization overhead of one task with respect to the schedule period for relevant wireless standards (since TD-SCDMA has non-uniform symbol length, we took the duration of the midamble – the shortest symbol – as the schedule period).

The overhead is negligible for broadcast and connectivity standards. For WLAN, an OFDM symbol arrives every 4 μs, and the overhead is 16.3% of the total schedule period. This overhead however, will decrease for higher processor clock frequencies, since the cycle count of the streaming kernel functions and the schedule period remain the same. As of the time of writing (2013), modems are using ARM processors with considerably higher clock speeds (>5 times faster) than used in this demonstrator, with larger caches and faster memory and bus architectures. Vector processors have also experienced a clock speed increase of more than 6×. The overhead of synchronization for faster architecture can be very conservatively considered to have decreased by a factor of 5 with respect to the scheduling period, which is fixed for each standard. It is also of note that in this demonstrator we made a very straightforward port of the SoD Streaming Kernel to the EVP, without using any EVP intrinsics, for instance, when dealing with computation of memory addresses to access the circular FIFO buffers.

8.3.4 Multi-radio Operation

The prototype can only employ NPNBRR schedulers on the EVPs since its version of the core has no interrupt support. Consequently, only radios working at about the same time granularity can share resources, e.g. 2×DVB-T or 2×WLAN. The DF Scheduler computed resource budgets that accommodate these use cases using 2 ARM clusters and 1 EVP cluster. In the WLAN we scaled the execution times by 2, due to the low processor clock speeds on the prototype board. Using an interrupt-enabled EVP, a TDM scheduler with a 4 μs period allows DVB-T and WLAN to run together. This would cost a 13.7% context switching overhead, which is not insignificant, but affordable for HRT multi-radio. The 2×WLAN use-case can only be mapped with NPRR schedulers when clustering is applied. In fact, clustering is essential for demanding standards such as WLAN, dramatically reducing synchronization overhead. For WLAN, without clustering, each actor adds at least the 16.3% overhead shown in Table 8.1 to the total resource expenditure, making mapping of two WLAN jobs sharing resources impossible. With clustering, such overhead is only incurred once per cluster, and the mapping of two WLAN becomes possible. Table 8.1 also shows that for standards other than WLAN the synchronization overhead is much more acceptable. Again, notice that these task synchronization overheads were obtained by simply compiling the SoD API C code on EVP, with no platform-specific optimizations, so there may still be plenty of room for optimization.

8.3.5 Resource Fragmentation

We have not experienced resource fragmentation after multiple start/stop requests. Our demonstrator supports few simultaneous transceivers and instances (4 of each). Since WLAN cannot share processors with slower radios, such as DVB-T or TD-SCDMA, due to the lack of preemptive schedulers in the demonstrator, there is a natural limitation to the mappings: WLANs can only be mapped to empty processors or to processors where other WLANs are already running. When a WLAN is stopped, it frees the exact space that a new WLAN instance will require.

8.4 Conclusion

The demonstrator shows how the framework concepts proposed in Chap. 2 can be put to practice, even on a platform that was originally designed as a demonstrator for single-radio operation. It also allows us to measure the memory and performance overhead that each one of our run-time components add to a modem platform. For our implementation, these overheads turned out to still allow for two WLAN receivers to run simultaneously for a processor with 2× the clock frequency of the EVPs in the prototype board – which were chosen to allow the execution of one single WLAN –, although with a high scheduling overhead of 16% of the WLAN period per task. Weighing in the fact that WLAN is a standard with a much faster rate of operation than most other radio standards – in fact, for popular connectivity standards this overhead is well bellow 1% of the cycle budget – and that the implementation of the inter-task communication primitives was done by re-compiling off-the-shelf C code, it is safe to state that the communication overhead of a more realistic implementation for a less demanding standard should be negligeable.

Chapter 9
Conclusions and Further Work

We live in a planet overflown with information, where billions of wirelessly connected devices constantly send and receive data to each other. Wireless connectivity is enabled through several different radio standards. Devices need to handle this diversity, and, at times, must support several of these standards simultaneously.

The baseband layer of a wireless communication standard is often implemented on a heterogeneous multiprocessor platform. Each wireless standard has its own strict temporal requirements. In systems that support a single standard at a time, the problem of mapping the application to the hardware with guaranteed temporal behavior is complex, but known to be manageable from an engineering perspective. This is typically done by carying out all mapping and scheduling decisions manually, or with little tool support, and exhaustively verifying all possible use cases.

However, in an environment where multiple such transceivers may be active simultaneously, the problem becomes considerably more complex, as resource contention among transceivers may lead to unpredictable temporal behavior.

A solution to this problem is to assign fully separated resources to each transceiver. Such a solution is simple and clean, but neglects the fact that sharing resources can lead to reduced costs in chip area, and cost is still a major driving factor in high volumes electronics.

In this book we have presented techniques to share hardware resources between transceivers executing simultaneously, while still being capable of giving hard real-time guarantees to each transceiver. It is important to note, however, that the techniques presented are not exclusively targeted towards the radio domain. In fact, they are designed to have in mind the requirements of a broader set of applications that can be classified as real-time streaming applications.

We described a software framework that is capable of handling multi-radio execution with real-time guarantees, if used in conjunction with a predictable hardware platform, including an automated programming flow and a multiprocessor runtime for resource management, scheduling and inter-task communication. Our departing point was the observation that the main sources of unpredictable timing

O. Moreira and H. Corporaal, *Scheduling Real-Time Streaming Applications onto an Embedded Multiprocessor*, Embedded Systems 24, DOI 10.1007/978-3-319-01246-9_9,

behavior in the execution of an application are the functionality of the application itself and the runtime resource management policies of the execution platform.

To address runtime issues, we used FIFO-based communication and synchronization, local budget schedulers per processor, and a global resource reservation mechanism that effectively isolates the worst-case temporal behavior of each transceiver from the behavior of other running transceivers.

As a basis for a strict model of computation with strong analytical properties and fitting expressivity, we selected data flow: its static variants have strong analytical properties, and since it is a concurrent, asynchronous model of computation, it lends itself well to distributed implementations.

We looked at how the a software framework and a runtime environment based on these concepts can be organized in Chap. 2. The usage of a high-level language to describe data flow guarantees correctness by construction and enables automatic model extraction and automatic code generation, which renders a more reliable tool chain, makes the source code more portable by allowing automatic generation of platform-specific API calls, and dramatically simplifies the job of the implementer.

In Chap. 3, we provided an overview of several existing flavors of data flow models of computation.

In Chap. 4, we looked at latency and throughput analysis of data flow. We have shown how we can guarantee a strictly periodic behavior for an actor in the graph, independently of any timing variations due to start-up behavior or variable, (but bounded) execution times of other actors in the graph, provided that an external source has a behavior that is lower-bounded by a periodic regime. We presented analysis techniques that allow us to verify maximum bounds on latency – and therefore provide maximum latency guarantees – for the self-timed execution of a multi-rate data flow graph, assuming either a periodic source, a bursty source or a sporadic source.

In Chap. 5, we discussed architectural modeling in data flow, and presented a scheduling strategy and an automated scheduling flow that enable the simultaneous execution of multiple hard-real-time data flow jobs. Each job has its own execution rate and starts and stops independently from other jobs, at instants unknown at compile-time, on a multiprocessor system-on-chip. We have shown how such jobs, if described by a Multi-Rate Data Flow graph, can be automatically mapped onto a multi-processor system.

The mapping strategy presented in this book differentiates between intra-job scheduling and inter-job scheduling. Intra-job scheduling is handled at compile-time, through the generation of clusters of statically-ordered actors, and the computation of buffer sizes for arcs between actors. Inter-job scheduling is solved at run-time by using schedulers such as either Time-Division Multiplex (TDM) or Non-Preemptive Non-Blocking Round Robin (NPNBRR), depending on whether the processor supports preemption or not. At compile time, we compute settings for the runtime schedulers per job.

We showed how a combination of TDM or NPNBRR and static-order scheduling can be modeled as additional nodes and edges on the data flow representation of the job using Single-Rate Data flow semantics, to enable tight worst-case temporal analysis. Based on this analysis model, we presented algorithms to find combined TDM/NPNBRR and static order schedules for jobs that guarantee a requested minimum throughput and maximum latency, while minimizing the usage of processing resources. We illustrated the usage of these techniques for a combination of Wireless LAN and TD-SCDMA radio jobs running on a prototype Software-Defined Radio platform.

One of the traditional limitations of data flow with respect to our preferred application domain is that transceivers do have some data dependent behavior which cannot be represented by the static variants of data flow (single-rate, multi-rate and cyclo-static) that allow temporal analysis. In Chap. 6 we presented an extension of static data flow that allows a limited amount of data dependent behavior, while preserving most of the temporal properties, called Mode-Controlled Data Flow (MCDF). We described timing analysis techniques that allow us to reason both in terms of minimum sustainable throughput and maximum latency for MCDF graphs, and mapping techniques that allows us to compute budgets for MCDF graphs, by extending the techniques discussed in Chap. 5.

Given the budgets computed by the compile-time tools, we need to allocate and reserve resources across the multiprocessor to meet the budget requirements of a transceiver, whenever the transceiver is required to execute. In Chap. 7, we described algorithms to solve the runtime mapping problem, including allocation of network resources in hardware architectures where a Network-On-Chip is used for inter-processor communication.

To bring all the components of this solution together, we made an implementation of the mapping tools and of the runtime software on a prototype chip for baseband processing. This experiment, described in Chap. 8, allowed us to show that the software framework could be built from a combination of standard runtime OSes, off-the-shelf compilation tools, and some custom software, in a modular manner. We showed that the run-time could execute on a pre-existing hardware platform, and that the runtime overhead required by the discussed scheduling policies was acceptable from the perspective of the application requirements.

Nonetheless, the list of topics that require future work is a long one. To refer only some of the most important ones:

- (Quasi)-Static-order scheduling and clustering algorithms: the algorithm for static ordering has exponential complexity, and can, for larger graphs, take a long time to terminate. Its adaptation to handle quasi-static ordering for MCDF suffers from even larger execution times, due to the fact that many different static orders correspond to the same quasi-static order (this is attenuated somehow by our usage of state hashing). The development of heuristics for static-ordering and quasi-static ordering, potentially combined with other steps of the mapping flow such as scheduler settings determination and buffer sizing minimization is certainly useful, and very likely necessary, if larger graphs are to be handled.
- MCDF Extensions: Mode-Controlled Data Flow, as presented in this book, is single-rate; although some limited forms of multi-rate behavior should be reasonably easy to handle with the existing techniques, it would be interesting

to study what gains in expressivity could come from allowing, for instances, multi-rate behavior for the variable-rate actors. Also, the imposition of a single mode controller makes temporal analysis much simpler, but makes it very cumbersome to express applications with a very complex control structure. This suggests that further work is needed in supporting MCDF graphs with multiple mode controllers, including the special case of hierarchical control.

- MCDF buffer sizing: in this book we only provide crude upper bounds to the amount of buffer sizing necessary to meet the timing requirements for a given MCDF graph, using pre-defined reference static-order schedules. There is work to do on optimized buffer sizing algorithms. A promising lead here is the work on buffer sizing for Variable-Phased Data Flow (VPDF).
- Budget Schedulers: in our work, we used only NPNBRR and TDM schedulers. TDM, however, provides rather poor trade-offs between response latency and throughput – for instance, if one wants a low latency for a task, one may need to allocate many more TDM resources than required by the load of the task, just to reduce arbitration time. Other schedulers, such as the Priority-based Budget Scheduler [82] or the Credit-Controlled Static Priority arbiter [2] allow for better resource usage. There is nothing in the proposed flow that cannot be easily adapted to work with other schedulers, as long as a conservative data flow model can be provided for the temporal behavior of a static-order of single-rate data flow actors running on such a scheduler – which is the case for both PBS and CCSP.
- Resource Fragmentation: The resource management framework presented in Chap. 7 assumes that memory does not fragment. The reason for this is that we assumed a resource defragmentation technique would be available. To the best of our knowledge, the problem is still largely unsolved.
- Power-aware scheduling was probably the most important aspect that we neglected in this book.For portable devices, which are powered by a small battery and cannot use the same sort of cooling technology found in more sedentary devices, power and energy consumption are very important, even primary, concerns.

All of the listed problems are of research interest, and the solutions that we will find for them will certainly impact the adoption of data flow techniques by developers of embedded RT streaming applications. It is our opinion that very satisfactory solutions can be found for most, probably all, of them.

However, the adoption by the industry of many of these concepts and technologies is challenged by other factors that are more difficult to overcome.

To start, there are the customary issues related with the adoption of a new technology. Most successful companies have mature design processes in place, and staffs of extremely gifted, experienced and knowledgeable engineers. A change to a platform design strategy centered on budget-based resource allocation and data-flow-based programming may be a step too far, as it requires new skills, new processes and new tools. Even if the current frameworks used in the industry cannot handle multi-application real-time requirements at all, it is likely that small incremental changes to the current approach, however complex or inefficient, will be preferred rather than a departure towards a completely new type of framework. So a challenge for data flow technologies is finding a likely incremental path to adoption.

This brings us to another issue: the technology is not yet mature. Not only there are still many problems to address, but, for the time being, we cannot offer mature tooling support for data flow-based system design. Tools such as the ones we developed during the course of our work, and similar ones, are not ready for the prime time. They have been designed primarily as research vehicles by small teams of researchers, not as industrial-strength CAD tools.

Even more challenging may be the fact that the complexity and even feasibility of data flow modeling of hardware architecture constraints is very dependent on the features of the platform. For instances, fixed priority arbiters and schedulers are difficult to model satisfactorily, but they are very common in existing platforms. Even if it is often true that many such architectural choices prevent the system from being predictable or composable, and should therefore not be used in designing a hard-real time system, it is also often true that although a given scheduler or arbiter has no bounded worst case in general, a bound can still be provided under certain reasonable assumptions. A good example of this is again a fixed-priority scheduler – once we have a best-case characterization of the activation pattern and resource usage of all higher priority tasks, we can easily provide a tight characterization of the temporal behavior of any given task. This breaks the basic principles of the approach proposed in this book, as it requires knowledge of the temporal behavior of one or more other jobs (including their best case behavior) in order to characterize the worst case behavior of another job, and it does not preserve temporal monotonicity, becoming as such vulnerable to scheduling anomalies. However, in practice, it may work for many specific cases. It is certainly possible that during the design process or even later, when the product is already delivered, it will fail due to, for example, a last minute change in the behavior of a high priority task that was not accounted for in the temporal analysis of other tasks.

Nonetheless, the fact remains that some form of analysis is possible, in ideal conditions, without the usage of the techniques described in this book, and this in turn makes it difficult to argue a priori that such system is not going to be predictable and that it must be completely changed to accommodate for a data flow-driven design, leaving the data flow advocate often in the role of a "prophet of doom", constantly warning developers against hypothetical future catastrophes.

The bottom line is that data flow based techniques must be introduced to the industry gradually. In fact, this is one of the reasons we strove to have a modular software framework in Chap. 2. Each component of the framework was designed such that it is useful by itself, even if the other components are missing. For instances, the resource manager can be combined with manual generation of budgets. The data flow analysis tools can be used to analyze a data flow model that was manually extracted from the application description. A data flow input language, combined with a good code generator, can provide a lot of value by allowing the generation of platform-independent code, and guaranteing that platform-specific APIs are correctly used. “Selling” the approach in bits is one way of building an incremental path to adoption.

It is of prime importance to understand the application, the requirements and the platform; understand that real-time behavior is one of the many concerns of the design team, accept that sometimes we will have to do with non-enforceable assumptions and try to make the analysis work with as few changes as possible to the platform and design process.

In any case, and however difficult the introduction of data flow may be, it may soon become absolutely indispensable for the industry to find a new way of designing modems, and other complex real-time streaming applications with similar characteristics. The techniques that are currently in use cannot scale with the increased complexity of the applications, and the increase in the number of combinations of applications that must be supported. We believe the approach presented in this book achieves a good compromise between simplicity, performance and robustness, while meeting the requirements, but it still needs the maturity that only years of actual product development can bring.

Bibliography

1. A. Ahtiainen et al. Multiradio scheduling and resource sharing on a software defined radio computing platform. In *Proc. Software-Defined Radio Forum Technical Conference*, Oct. 2008.
2. B. Akesson, L. Steffens, E. Strooisma, and K. Goossens. Real-time scheduling using credit-controlled static-priority arbitration. In *Proceedings of the 2008 14th IEEE International Conference on Embedded and Real-Time Computing Systems and Applications*, pages 3–14, Washington, DC, USA, 2008. IEEE Computer Society.
3. F. Baccelli, G. Cohen, G. Olsder, and J.-P. Quadrat. *Synchronization and Linearity*. John Wiley & Sons, Inc., 1992.
4. S. K. Baruah and S. Funk. Task assignment in heterogenous multiprocessor platforms. Technical report, University of North Carolina, 2003.
5. J. Beck and D. Siewiorek. Modeling multicomputer task allocation as a vector packing problem. In *Proceedings of the 9th international symposium on System synthesis*, ISSS '96, pages 115–, Washington, DC, USA, 1996. IEEE Computer Society.
6. M. Bekooij et al. Predictable embedded multiprocessor system design. In *Proc. Int'l Workshop SCOPES*, LNCS 3199. Springer, Sept. 2004.
7. M. Bekooij et al. Dataflow analysis for real-time embedded multiprocessor system design. In *Dynamic and Robust Streaming in and between Connected Consumer Electronic Devices*, volume 3, pages 81–108. Springer, 2005.
8. K. Berkel et al. Vector processing as an enabler for software-defined radio in handheld devices. *EURASIP Journal on Applied Signal Processing*, 2005(16), 2005.
9. G. Bilsen et al. Cyclo-static dataflow. *IEEE Transactions on Signal Processing*, 44(2):397–408, 1996.
10. J. Buck. *Scheduling dynamic dataflow graphs with bounded memory using the token flow model*. PhD thesis, Univ. of California, Berkeley, September 1993.
11. J. Buck. Static scheduling and code generation from dynamic dataflow graphs with integer-valued control streams. In *Proc. of the Asilomar Conference on Signals, Systems, and Computers*, 1994.
12. G. Buttazzo. *Hard Real-Time Computing Systems*. Kluwer Academic Publishers, 1997.
13. J. Cochet-Terrasson et al. Numerical computation of spectral elements in max-plus algebra. In *Proc. IFAC Conf. on Syst. Structure and Control*, 1998.
14. E. G. Coffman, Jr., E. G. Coffman, D. S. Johnson, D. S. Johnson, L. A. Mcgeoch, L. A. Mcgeoch, R. R. Weber, and R. R. Weber. Bin packing with discrete item sizes part ii: Average-case behavior of ffd and bfd. *In preparation*, 13:384–402, 1997.
15. T. Corman et al. *Introduction to Algorithms*. McGraw-Hill, 2001.
16. D. Culler et al. *Parallel Computer Architecture: a hardware/software approach*. Morgan Kaufmann, 1999.

O. Moreira and H. Corporaal, *Scheduling Real-Time Streaming Applications onto an Embedded Multiprocessor*, Embedded Systems 24, DOI 10.1007/978-3-319-01246-9,

17. A. Dasdan. Experimental analysis of the fastest optimum cycle ratio and mean algorithms. *ACM Transactions on Design Automation of Electronic Systems*, 9(4):385–418, Oct. 2004.
18. E. W. Dijkstra. Hierarchical ordering of sequential processes. *Acta Informatica*, pages 115–138, 1971.
19. D. Feitelson. Job scheduling in multiprogrammed parallel systems. Technical report, IBM Research Report RC, 1994.
20. M. R. Garey and D. S. Johnson. *Computers and Intractability; A Guide to the Theory of NP-Completeness*. W. H. Freeman & Co., New York, NY, USA, 1990.
21. M. Geilen. Dataflow scenarios. In *IEEE Transactions on Computers*, 2010.
22. M. Geilen and S. Stuijk. Worst-case performance analysis of synchronous dataflow scenarios. In *ACM CODES + ISSS*, 2010.
23. A. Ghamarian, M. Geilen, S. Stuijk, T. Basten, A. Moonen, M. Bekooij, B. Theelen, and M. Mousavi. Throughput analysis of synchronous data flow graphs. In *Proc. Int'l Conf. on Application of Concurrency to System Design (ACSD)*, pages 25–34, June 2006.
24. A. H. Ghamarian, S. Stuijk, T. Basten, M. Geilen, and B. D. Theelen. Latency minimization for synchronous data flow graphs. In *DSD*, pages 189–196, 2007.
25. A. Girault, B. Lee, and E. Lee. Hierarchical finite state machines with multiple concurrency models. *IEEE Trans. on Computer-Aided Design of Integrated Circuits and Systems*, 18(6):742–760, June 1999.
26. GNU Project. GLPK website. In http://www.gnu.org/software/glpk/, 2011.
27. S. Goddard and K. Jeffay. Managing latency and buffer requirements in processing graph chains. *The Computer Journal*, 44(6), 2001.
28. K. Goossens, J. Dielissen, J. van Meerbergen, P. Poplavko, A. Rădulescu, E. Rijpkema, E. Waterlander, and P. Wielage. *Guaranteeing the quality of services in networks on chip*, pages 61–82. Kluwer Academic Publishers, Hingham, MA, USA, 2003.
29. K. Goossens, P. Wielage, A. Peeters, and J. Van Meerbergen. Networks on silicon: Combining best-effort and guaranteed services. In *Proceedings of the conference on Design, automation and test in Europe*, DATE '02, pages 423–, Washington, DC, USA, 2002. IEEE Computer Society.
30. K. Goossens et al. *Guaranteeing the quality of service in networks on chip*. Kluwer, 2003.
31. R. Govindarajan, G. Gao, and P. Desai. Minimizing memory requirements in rate-optimal schedules. In *Proc. Int'l Conf. on Application-Specific Array Processors*, pages 75–86, Aug. 1993.
32. R. Govindarajan and G. G.R. A novel framework for multirate scheduling. In *International Conference on Application Specific Array Processors*, 1993.
33. M. Grant and S. Boyd. CVX: Matlab software for disciplined convex programming, version 1.21. http://cvxr.com/cvx, Feb. 2011.
34. M. Grant, S. Boyd, and Y. Ye. Disciplined convex programming. In *Global Optimization: From Theory to Implementation, Nonconvex Optimization and Its Application Series*, pages 155–210. Springer, 2006.
35. C. Grassmann, M. Richter, and M. Sauermann. Mapping the physical layer of radio standards to multiprocessor architectures. In *Proc. Design, Automation and Test in Europe Conference and Exhibition (DATE)*, pages 1412–1417, 2007.
36. W. Gropp et al. *Using MPI: Portable Parallel Programming with the Message Passing Interface*. MIT Press, 1994.
37. S. Ha and E. Lee. Compile-time scheduling of dynamic constructs in dataflow program graphs. *IEEE Transactions on Computers*, 46(7):768–778, July 1997.
38. A. Hansson. *A Composable and Predictable On-Chip Interconnect*. PhD thesis, Eindhoven University of Technology, 2009.
39. A. Hansson, K. Goossens, and A. Radulescu. A unified approach to constraint mapping and routing on network-on-chip architectures. In *Int'l Symposium on System Synthesis (ISSS)*, pages 75–80, 2005.

40. A. Hansson, K. Gossens, M. Bekooij, and J. Huisken. Compsoc: A template for composable and predictable multi-processor system on chips. *ACM Transactions on Design Automation of Electronic Systems*, Oct. 2009.
41. B. Hayes. Cloud computing. *ACM Communications of the ACM*, 51(7):9–11, July 2008.
42. W. Heisenberg. *The Physical Principles of Quantum Theory*. University of Chicago Press, 1930.
43. T. A. Henzinger. Two challenges in embedded systems design: predictability and robustness. *Philosophical Transactions of the Royal Society*, Oct. 2008.
44. International Telecommunications Union. The World in 2013 - Facts and Figures. http://www.itu.int, 2013.
45. M. Jersak, K. Richter, and R. Ernst. Performance analysis of complex embedded systems. *International Journal of Embedded Systems*, 1(1–2):33–49, 2005.
46. Joint Tactical Radio Systems Program Office. Software communicatons architecture specification. http://sca.jpeojtrs.mil/.
47. Joyent. What is cloud computing? www.youtube.com, 2008. 6PNuQHUiV3Q.
48. G. Kahn. The semantics of a simple language for parallel programming. In *Proceedings IFIP Congress*, pages 471–475, 1974.
49. B. Kernighan and S. Lin. An efficient heuristic procedure for partitioning graphs. *The Bell System Technical Journal*, 1970.
50. S. G. Kolliopoulos and C. Stein. Approximating disjoint-path problems using greedy algorithms and packing integer programs. In *Lecture Notes in Computer Science*, volume 1412/1998, pages 153–168. Springer-Verlag, 1998.
51. Kopetz.*Real-Time Systems: Design Principles for Distributed Embedded Applications*. Springer, 1997.
52. B. Korte and J. Vygen. *Combinatorial Optimization: Theory and Algorithms*. Springer-Verlag Berlin Heidelberg, 2002.
53. L. Kou and G. Markowski. Multidimensional bin packing algortihms. *IBM Journal of Research and Development*, 22(2), 1977.
54. P. Kourzanov. http://bitbucket.org/pjotr/lime/src/tip/doc/.
55. E. Lee. The problem with threads. *IEEE Computer*, pages 60–67, 2006.
56. E. Lee and D. Messerschmitt. Static scheduling of synchronous data flow programs for digital signal processing. In *IEEE Transactions on Computers*, 1987.
57. E. Lee and D. Messerschmitt. Synchronous data flow. In *Proceedings of the IEEE*, 1987.
58. A. Lele, O. Moreira, and P. J. Cuijpers. A new data flow analysis model for tdm. In *Proceedings of the tenth ACM international conference on Embedded software*, EMSOFT '12, pages 237–246, New York, NY, USA, 2012. ACM.
59. J. Liebeherr, A. Burchard, Y. Oh, and S. H. Son. New strategies for assigning real-time tasks to multiprocessor systems. *IEEE Trans. Comput.*, 44:1429–1442, December 1995.
60. Y. Lin, Y. L. Hyunseok, M. Woh, Y. Harel, S. Mahlke, T. Mudge, and C. Chakrabarti. Soda: A low-power architecture for software radio. In *In Proc. of the 33rd Annual International Symposium on Computer Architecture*, pages 89–101, 2006.
61. Y. Lin et al. Spex: A programming language for software defined radio. In *Proc. Software-Defined Radio Forum Technical Conference*, Oct. 2006.
62. Y. Lin et al. Hierarchical coarse-grained stream compilation for software defined radio. In *Proc. Int'l Conf. on Compilers, Architectures and Synthesis for Embedded Systems (CASES)*, October 2007.
63. C. L. Liu and J. W. Layland. Scheduling algorithms for multiprogramming in a hard-real-time environment. *J. ACM*, 20:46–61, January 1973.
64. T. Lundqvist and P. Stenström. Timing anomalies in dynamically scheduled microprocessors. In *Proc. Real-Time Systems Symposium*, 1999.
65. F. Malerba, R. Nelson, L. Orsenigo, S. Winter, F. Malerba, R. Nelson, L. Orsenigo, and S. Winter. A history-friendly model of the co-evolution of the computer and semiconductors industries: Capabilities and technical change as determinants of the vertical scope of firms in related industries, 2006.

66. T. Marescaux, B. Bricke, P. Debacker, V. Nollet, and H. Corporaal. Dynamic time-slot allocation for qos enabled networks on chip. In *ESTImedia*, pages 47–52, 2005.
67. G. Martin and H. Chang. *Winning the SoC Revolution*. Kluwer Academic Publishers, 2003.
68. J. Mitola. The software radio. In *Proc. IEEE National Telesystems Conference*. IEEE Computer Society Press, 1992.
69. T. Miyazaki and E. Lee. Code generation by using integer-controlled data flow graph. In *Proc. ICASSP*, 1997.
70. J. Mol. Resource allocation for streaming applications in multiprocessors. Technical report, Delft University of Technology, 2004.
71. A. Moonen, M. Bekooij, and J. van Meerbergen. Timing analysis model for network based multiprocessor systems. In *Proc. Workshop of Circuits, System and Signal Processing (ProRISC)*, pages 91–99, Veldhoven, The Netherlands, 2004.
72. O. Moreira, T. Basten, M. Geilen, and S. Stuijk. Buffer sizing for rate-optimal single-rate dataflow revisited. In *IEEE Transactions on Computers*, 2010.
73. O. Moreira, M. Bekooij, and J. Mol. Online resource management for a multiprocessor with a network-on-chip. In *Proc. ACM Symposium on Applied Computing*, March 2007.
74. O. Moreira, M. Bekooij, J. Mol, and J. van Meerbergen. Multiprocessor resource allocation for hard real-time streaming with a dynamic job mix. In *Proc. IEEE Real-Time and Embedded Technology and Applications Symposium (RTAS)*, March 2005.
75. O. Moreira, F. Valente, and M. Bekooij. Scheduling multiple independent hard-real-time jobs on a heterogeneous multiprocessor. In *Proc. Embedded Software Conference (EMSOFT)*, October 2007.
76. Q. Ning and G. Gao. A novel framework of register allocation for software pipelining. In *Proceedings of the ACM SIGPLAN-SIGACT Symposium on Principles of Programming Languages*, pages 29–42, 1993.
77. T. M. Parks. *Bounded Scheduling of Process Networks*. PhD thesis, Princeton University, 1987.
78. P. Poplavko, T. Basten, M. Bekooij, J. van Meerbergen, and B. Mesman. Task-level timing models for guaranteed performance in multiprocessor networks-on-chip. In *Proc. Int'l Conf. on Compilers, Architectures and Synthesis for Embedded Systems (CASES)*, pages 63–72, 2003.
79. U. Ramacher. Software-defined radio prospects for multi-standard mobile phones. *IEEE Computer*, 2007.
80. R. Reiter. Scheduling parallel computations. *Journal of the ACM*, 15(4):590–599, October 1968.
81. S. Sriram and S. Bhattacharyya. *Embedded Multiprocessors: Scheduling and Synchronization*. Marcel Dekker Inc., 2000.
82. M. Steine, M. Bekooij, and M. Wiggers. A priority-based budget scheduler with conservative dataflow model. In *DSD*, pages 37–44, 2009.
83. D. Stiliadis and A. Varma. Latency-rate servers: a general model for analysis of traffic scheduling algorithms. *IEEE/ACM Trans. Netw.*, 6(5):611–624, 1998.
84. M. Strik et al. Heterogeneous multiprocessor for the management of real-time video and graphics streams. *IEEE Journal of Solid-State Circuits*, 35(11):1722–1731, 2000.
85. S. Stuijk, M. Geilen, and T. Basten. A predictable multi-processor design flow for streaming applications with dynamic behavior. In *DSD*, 2010.
86. S. Stuijk et al. Multiprocessor resource allocation for throughput-constrained synchronous dataflow graphs. In *Proc. Design Automation Conference (DAC)*, 2007.
87. B. Theelen et al. Scenario-aware data flow model for combined long-run average and worst-case performance analysis. *Int'l Conf MEMOCODE*, 2006.
88. W. Thies. *Language and Compiler Support for Stream Programs*. PhD thesis, Massachusetts Institute of Technology, 2009.
89. D. tzen Peng, K. G. Shin, and T. F. Abdelzaher. Assignment and scheduling communicating periodic tasks in distributed real-time systems. *IEEE Transactions on Software Engineering*, 23:745–758, 1997.
90. M. Wiggers. *Aperiodic Multiprocessor Scheduling*. PhD thesis, University of Twente, June 2009.

91. M. Wiggers, M. Bekooij, M. Geilen, and T. Basten. Simultaneous budget and buffer size computation for throughput-constrained task graphs. In *Proceedings of the Design, Automation and Test in Europe (DATE) Conference*, 2010.
92. M. Woh et al. From soda to scotch: The evolution of a wireless baseband processor. In *ACM International Symposium on Microarchitecture*, 2008.
93. O. Zapata and P. Mejia-Alvarez. Analysis of real-time multiprocessors scheduling algorithms. In *Proceedings of the Real-Time Systems Symposium (RTSS)*, 2003.